The Uses of Place-Names

Other books from Scottish Cultural Press

Scotland in Dark Age Britain
Barbara Crawford (ed.)
1 898218 61 7

Scottish Woodland History
T.C. Smout (ed.)
1 898218 53 6

The History of Soils and Field Systems
S. Foster and T.C. Smout (eds.)
1 898218 13 7

Scotland since Prehistory: Natural Change and Human Impact
T.C. Smout
1 898218 03 X

Fragile Environments: The Use and Management of Tentsmuir NNR
Graeme Whittington (ed.)
1 898218 77 3

Scotland's Rural Land Use Agencies: The history and effectiveness in Scotland of the Forestry Commission, Nature Conservancy Council and Countryside Commission
Donald G. Mackay
1 898218 31 5

The Uses of Place-Names

Edited by
Simon Taylor

St John's House Papers No. 7
St Andrews

SCOTTISH CULTURAL PRESS
EDINBURGH

First Published 1998 for
The St Andrews Scottish Studies Institute
by

Scottish Cultural Press
Unit 14, Leith Walk Business Centre
130 Leith Walk
Edinburgh EH6 5DT
Tel: 0131 555 5950 • Fax: 0131 555 5018
e-mail: scp@sol.co.uk

Copyright © 1998 St Andrews Scottish Studies Institute on behalf of the contributors

All rights reserved. No part of this publication may be reproduced, stored in a retrieval system, or transmitted in any form or by any means, electronic, mechanical, photocopying, recording or otherwise without the prior permission of Scottish Cultural Press.

British Library Cataloguing in Publication Data
A catalogue record for this book is available from the British Library

ISBN: 1 898218 98 6

Printed and bound by
Cromwell Press, Trowbridge, Wiltshire

Contents

	List of Illustrations	vii
	Contributors	ix
	Foreword *I.A. Fraser*	xi
	Acknowledgements	xiii
	County Abbreviations	xiv
	Introduction *Simon Taylor*	1
1	Place-names as a Resource for the Historical Linguist *Roibeard Ó Maolalaigh*	12
2	The Uses of Place-names and Scottish History – pointers and pitfalls *G.W.S. Barrow*	54
3	Place-names and Landscape *Margaret Gelling*	75
4	Place-name Distributions and Field Archaeology in South-west Wales *Terrence James*	101
5	Scandinavian Settlement in Unst, Shetland: archaeology and place-names *Steffen Stummann Hansen & Doreen Waugh*	120
6	Place-names and Literature: Evidence from the Gaelic Ballads *Donald E. Meek*	147
7	Gwaun Henllan – the Oldest Recorded Meadow in Wales? *Heather James*	169
	Afterword *W.F.H. Nicolaisen*	180
	Index	182

List of Illustrations

2.1	Distribution of *aber*- and *inver*- place-names.	57
2.2	Distribution of *nemeton*- place-names, with other localities associated with pagan worship.	58
2.3	Distribution of *pol*- place-names.	60
2.4	Central Scottish Borders, showing *pol*- place-names.	61
2.5	Distribution of Ochiltree, Ogilface and Ogilvie.	63
2.6	Distribution of *pebyll*- place-names.	64
2.7	Place-names containing *monadh, munið*.	66
2.8	Place-names containing O.E. *burh/burg* in southern Scotland and northern England.	69
2.9	Place-names containing O.E. *boþl* in Southern Scotland and Northern England.	69
2.10	Place-names containing Older Scots *threep*- 'debateable', showing the different generic elements with which it is combined.	71
3.1	Buckinghamshire *dūn* country.	76
3.2	The Chilterns and the Vale of Aylesbury, showing parts of Berkshire, Buckinghamshire, Oxfordshire, Bedfordshire, Hertfordshire and Middlesex.	77
3.3a–d	Illustrations of places with names containing Anglo-Saxon *dūn*.	79
3.4a–d	Illustrations of places with names containing Anglo-Saxon *beorg* or Old Norse *berg*.	80
3.5a–d	Illustrations of places with names containing Welsh *crūc*.	82
3.6a–d	Illustrations of places with names containing Anglo-Saxon *hōh*.	83
3.7a–d	Illustrations of places with names containing Anglo-Saxon **ofer*.	84
3.8	Illustration of a place with a name containing Anglo-Saxon *ōra*.	85
3.9	A view of Pershore, from Pensham Hill, from T. Nash's *Collections for the History of Worcestershire*, 1799.	86
3.10	Distribution of *ōra*- and **ofer*- place-names in England, showing ancient travel routes.	87
3.11	*Ōra*- and **ofer*- place-names in the West Midlands showing the Droitwich Salt Ways.	88
3.12	Relative distribution of *cumb* and *denu* place-names in the Chilterns, South-east England.	88
3.13a–c	Illustrations of places with names containing Anglo-Saxon *denu*.	89
3.14a–b	Illustrations of places with names containing Anglo-Saxon *cumb*.	90
3.15	Aerial photograph of Hope Dale SHR, looking north.	91
3.16	Map of Hope Dale SHR.	92
3.17	Illustration of a place with a name containing Anglo-Saxon *hop*.	92
3.18a–b	Illustrations of places with names containing Anglo-Saxon *ēg*.	93
3.19	Buildwas SHR, with the River Severn in flood.	94
3.20	Bleadney SOM: probable drainage system in the Anglo-Saxon period.	95
3.21	Place-names containing Anglo-Saxon *hȳth* in England.	96
3.22	Illustration of a place with a name containing Anglo-Saxon *pæth*.	97
4.1	*caer/gaer*-named hillforts in Dyfed, South-west Wales.	102

4.2	Carew Castle, Carew parish PEM (SN044037), with crop marks showing evidence of the former prehistoric promontory fort in the foreground.	104
4.3	Distribution of *castell* and *castle* place-names in Dyfed.	105
4.4	Distribution of *din* and *dinas* place-names in Dyfed.	105
4.5	*rath* in Pembrokeshire showing medieval Lordships.	107
4.6	*-ton* in Pembrokeshire showing medieval Lordships.	107
4.7	Hillforts in Dyfed not named *caer/castell/castle/rath*.	108
4.8	Distribution of 'castle' place-names in Dyfed.	109
4.9	The site of Castell Draenog, Llanboidy parish CRM (SN213214), showing as a bi-vallate cropmark in grassland.	110
4.10	*Castell/castle* hillforts in Dyfed close to churches and chapels.	111
4.11	Dyfed, showing commotes.	112
4.12	Lan, Llanboidy parish CRM (SN216205).	114
4.13	*sarn* names in Carmarthenshire.	116
4.14	Distribution of *ynys* place-names in Dyfed.	117
4.15	SMR sites named *parc* and *cae*.	118
5.1	Peter Moar (1888–1983) photographed standing in entrance of his place of work, the Danish Dairy, Commercial Street, Lerwick, in approx. 1911.	124
5.2	View from the south west over Haroldswick, one of the main settlements in Unst.	125
5.3	Distribution map of Norse sites in Unst.	126
5.4	Norse longhouse at Hamar.	127
5.5	Plan of Norse site at Hamar.	127
5.6	Plan of Norse site at Gardie I (Brookpoint).	129
5.7	View over the Sandwick-North site during excavation.	131
5.8	Ornamented bone-comb with copper rivets from the Sandwick-North site.	133
5.9	Norse longhouses of the ninth and tenth century.	135
5.10	Typological development of Viking and Early Medieval architecture in Denmark.	137
6.1	Irish locations associated with Fraoch traditions.	152
6.2	Place-names in Connacht connected with the *Laoidh Fhraoich* ('The Lay of Fraoch').	152
6.3	Locations of *Beann Ghulba(i)nn* in Ireland and Scotland.	155
6.4a	Benbulben, Co. Sligo.	156
6.4b	*Beann Ghulbainn*, Glenshee PER.	156
6.4c	*Beinn Ghuilbin*, north of Aviemore.	157
6.4d	Map showing exact position of *Beinn Ghuilbin* north of Aviemore.	157
6.5	Identifiable place-names in 'Caoilte and the Creatures'.	160
6.6a	Loch Freuchie, showing the wooded island near the south shore of the loch.	163
6.6b	The location of Loch Freuchie.	163
7.1	Location map of Llandybïe, Carmarthenshire.	170
7.2	The medieval parish of Llandybïe, with the bounds of *Maenor Meddynfych*.	172
7.3	The parish of Llandybïe.	174

Contributors

Geoffrey W.S. Barrow, former Sir William Fraser Professor of Scottish History and Palaeography, now Honorary Research Fellow, Department of Scottish History, University of Edinburgh.

Ian A. Fraser, Senior Lecturer and Director of the Scottish Place-Name Survey, School of Scottish Studies, University of Edinburgh.

Margaret Gelling, Honorary Reader in Place-Name Studies, University of Birmingham.

Heather James, Archaeological Officer, Dyfed Archaeological Trust/Ymddiriedolaeth Archaeolegol Dyfed, Llandeilo, Carmarthenshire.

Terrence James, Head of Information Services, Royal Commission on the Ancient and Historical Monuments of Wales/Comisiwn Brenhinol Henebion Cymru, Aberystwyth.

Donald E. Meek, Professor and Head of the Department of Celtic, University of Aberdeen.

Roibeard Ó Maolalaigh, Lecturer in Celtic Studies, Department of Celtic, and Director of Ionad na Gaeilge/Centre for Irish Studies, University of Edinburgh;

W.F.H. Nicolaisen, Distinguished Professor emeritus of English and Folklore, State University of New York at Binghamton, USA and Honorary Research Fellow, Department of English, University of Aberdeen.

Steffen Stummann Hansen, Research Lecturer, Institut for Arkæologi og Etnologi/Institute of Archaeology & Ethnology, University of Copenhagen, Denmark.

Simon Taylor, Anderson Research Fellow, St Andrews Scottish Studies Institute, University of St Andrews.

Doreen J. Waugh, Assistant Head, The Mary Erskine School, Edinburgh and Honorary Research Fellow, Department of English Language, University of Glasgow.

Foreword

From the conference opening remarks by Ian A. Fraser, director of the Scottish Place-Name Survey, School of Scottish Studies, University of Edinburgh.

It is with great pleasure that I welcome you to this conference, which is an important event in the development of Scottish place-name studies. It is particularly gratifying that so many people from all walks of life and from many academic disciplines have expressed their interest in Scottish toponymics by being present here today. And I extend an especially warm welcome to those of you who have made the journey from furth of Scotland.

A few weeks ago I received in my mail an off-print from a learned journal. There is nothing remarkable in that, since scarcely a week goes by but that we in academia receive such publications, solicited or otherwise. However, this seemed somewhat outwith my field, as it was an extract from *Mammal Review* 1995, Volume 25, no. 4. The title was 'Place-name evidence for the former distribution and status of Wolves and Beavers in Britain', by C. Aybes and D.W. Yalden, of the School of Biological Sciences, University of Manchester. According to the abstract the article attempted to provide a distribution of wolves and beavers using place-name evidence from a wide variety of printed sources, as well as including similar material for the incidence of brown bears, wild boar and wild cats. Clearly the authors had enlisted expert assistance, including such distinguished toponymists as Dr Margaret Gelling, one of today's speakers.

Such forays into toponymics remind us of the interdisciplinary nature of our subject. If place-name studies are to be of benefit to scholarship in the fields of archaeology, history and language we must also realise their importance as tools to scholars in the wider academic field. Our colleagues in other countries have made great strides in research and publication, and the work carried on in England, Northern Ireland and the Republic of Ireland, who all have well-established surveys, has resulted in name-studies being brought to the notice of the wider public, as well as being of immense value to the academic community.

Later on today, after the conference, we hope to establish the Scottish Place-Name Society, in an attempt to bring together all those with an interest in Scottish place-names. The Society will be open to all, with the expectation that it will be a forum for discussion, as well as being a force for the development of Scottish toponymics at all levels. More importantly it will help co-ordinate the wealth of expertise and experience which individual researchers have accumulated over the years. Recognising Scotland's complex toponymic history, we require input from many disciplines – linguistics, history, geography, archaeology, ethnology and many others.

We are also conscious of our links with other countries: England, Ireland and Wales are the most obvious, but there has been a long history of involvement with Scandinavian scholars, and a continuing interest in the Norse contacts with Scotland, which has resulted in a number of valuable recent studies.

One of the primary objectives of those closely involved in the study of Scottish place-names is the establishing of a Database for all Scottish place-names. This is urgently needed, as it will provide all the necessary information in a readily available format, and will be of immense value to researchers in many fields. If the publication of place-name material is to be satisfactorily achieved, the Database is vital, and we are already in close contact with scholars in Wales and Ireland who are in the process of setting up similar facilities. It is to be hoped that we in Scotland will learn a great deal from their experience. This project will require substantial funding, and we are actively engaged in fund-raising at the moment.

Finally, may I thank you for expressing your interest in Scottish toponymics by attending today's conference in such large numbers. This is a real encouragement for us, and we are very grateful for the support we have had from many individuals and organisations.

NOTE: The **Scottish Place-Name Society** was indeed founded on the evening of the conference. It has attracted a great deal of interest and support with membership currently (September 1997) at around 250.

For membership details, please contact the Scottish Place-Name Society c/o The School of Scottish Studies, 27 George Square, University of Edinburgh, Edinburgh EH8 9LD.

Acknowledgements

Apart from all the contributors to this volume, who have been models of promptness, co-operation and kindness with this first-time editor, I would like to thank in particular Dr Barbara Crawford for her valuable advice and support, as well as Dr Yvonne Burgess, Ms Dot Clark and Professor W.F.H. Nicolaisen for their comments on the Introduction; also Ms Carol Rodger of Scottish Cultural Press for her careful attention, and Mr Jim Renny for drawing many of the maps. Thanks are also due to the St Andrews Scottish Studies Institute for shouldering the main financial burden of production, as well as to a kind donor, who wishes to remain anonymous, for a contribution to the costs of the cover. Finally I have to thank the Anderson Research Fellowship, which has supported me during the preparation of this book.

County Abbreviations

(Pre-1975 counties)

England
BDF	Bedfordshire
BUC	Buckinghamshire
CAM	Cambridgeshire
CHE	Cheshire
DOR	Dorset
GLO	Gloucestershire
HNT	Huntingdonshire
LIN	Lincolnshire
LNC	Lancashire
NTB	Northumberland
NTP	Northamptonshire, England
OXF	Oxfordshire
SHR	Shropshire
SOM	Somerset
SSX	Sussex
SUR	Surrey
WAR	Warwickshire
WLT	Wiltshire
WML	Westmorland
WOR	Worcestershire
YON	Yorkshire, North Riding

Ireland
ANT	Antrim, Northern Ireland
ARM	Armagh, Northern Ireland
CLE	Clare
CLW	Carlow
CRK	Cork
CVN	Cavan
DNL	Donegal
DUB	Dublin
DWN	Down, Northern Ireland
FER	Fermanagh, Northern Ireland
GLY	Galway
KLD	Kildare
KLK	Kilkenny
KRY	Kerry
LFD	Longford
LMK	Limerick
LTH	Louth
MON	Monaghan
MTH	Meath
MYO	Mayo
OFY	Offaly
RSC	Roscommon
SGO	Sligo
TPY	Tipperary
TYE	Tyrone, Northern Ireland
WCW	Wicklow
WFD	Waterford
WMH	Westmeath
WXF	Wexford
WCW	Wicklow

Scotland
ABD	Aberdeen
ANG	Angus
BWK	Berwickshire
DNB	Dunbartonshire
DMF	Dumfriesshire
ELO	East Lothian
FIF	Fife
INV	Inverness-shire
KCD	Kincardineshire
LAN	Lanarkshire, Scotland
MLO	Midlothian
PEB	Peeblesshire
PER	Perthshire
ROS	Ross and Cromarty
ROX	Roxburghshire
STL	Stirlingshire
SUT	Sutherland
WIG	Wigtonshire

Wales
CRM	Carmarthenshire
PEM	Pembrokeshire

Introduction

The most basic use of a place-name is not in fact addressed in this book: place-names are there to help us find our way in the world. We would be, quite literally, lost without them. Most of the names on a Scottish map are 'just' names – they have no meaning beyond the fact that they refer to a feature in the landscape, be it natural or artificial. In the words of W.F.H. Nicolaisen, they tend to 'denote', i.e. individualise and exclude, rather than 'connote', i.e. permit abstraction and include (Nicolaisen 1988, 22–3). However, this was not always the case, as most of these names arose as descriptions which were understandable *as descriptions* to the local inhabitants – they both denoted *and* connoted.

Messages are therefore encoded in a name, messages which bear invaluable information about the name-givers and their world. Since many Scottish place-names were coined between 1,000 and 1,500 years ago, they are messages from a world which is otherwise extremely scantily recorded. Furthermore, the words used in the message, and the way they are put together, tell us more about the languages of Scotland before 1000 AD than any other source.

They are thus a very valuable linguistic and historical resource – in short, they are a precious part of our culture.

It was with this in mind that a conference was held in St Andrews in February 1996, entitled 'The Uses of Place-Names', which had as its theme an exploration of place-names as tools for a range of academic disciplines. Not only has this conference supplied the title for this book, but the conference papers, some of them much expanded, form most of its contents. Although held in Scotland, the speakers came from almost *aa the airts*, and the resulting book reflects this wide geographical spread, one which well matches its wide range of disciplines and themes: Scotland and Ireland provide the subject matter for Chapters 1 and 6; while Scotland alone is the subject of Chapter 2, and Shetland of Chapter 5, England is the focus of Chapter 3, and Wales of Chapters 4 and 7. This Introduction, however, is written very much from a Scottish perspective, and considers some of the applications and implications of both the Scottish and the non-Scottish material contained in these chapters for Scottish toponymics (place-name studies).

Introduction

Language

We start with language, which must be the starting point of any place-name work. For unless we know the language in which a place-name was coined, and can at least guess at its meaning or significance, then all the rest falls. There are at least seven languages which have left their mark on Scotland's place-names: Gaelic, Pictish, Norse, Welsh, Scots, English and French. Not all these languages are represented in every part of Scotland – for example, no Gaelic was ever spoken in the Northern Isles, and no Welsh was spoken on Skye. However, in any given area several of these languages are represented, and some would have existed at the same time. In **Chapter 1** Roibeard Ó Maolalaigh takes one language, Gaelic, and explores the contribution which place-names can make to our understanding of its early development in Scotland. Never before have Scottish Gaelic place-names been subjected to such detailed linguistic analysis, and not only are several important conclusions reached, but also the direction is set for much future research. Furthermore, the way that different languages interact with each other is always an important consideration, especially within a Scottish context, and is one that we ignore at our peril. Ó Maolalaigh also tackles this question, examining the transmission from Gaelic into Scots, and showing that the analysis of such transmission can teach us important things about both the donor and the host language.

History

The use of place-names by historians and historical geographers must perforce overlap. Professor Geoffrey Barrow has, during his distinguished career as one of Scotland's foremost medieval historians, made frequent use of place-names in various ways: in connection with early administrative units, in his chapter 'Pre-Feudal Scotland: Shires and Thanes' (Barrow 1973, 7–68); with medieval agrarian and social organisation, in 'Rural Settlement in central and eastern Scotland' (*ibid*. 257–78); as well as with specific aspects of such, in 'Popular Courts in Early Medieval Scotland: Some Suggested Place-Name Evidence', (Barrow 1981); with the early Christianisation of Pictland in 'The Childhood of Scottish Christianity: a Note on Some Place-Name Evidence' (Barrow 1983); and with medieval roads and travel in 'Land Routes: The Medieval Evidence' (Barrow 1984).

In **Chapter 2** Barrow, after a useful survey of toponymics in Scottish historiography, takes a variety of elements from different languages and looks at their distribution with a historian's eye, relating place-name evidence not only to known historical movements of peoples, but also to other aspects of early society such as religion and pastoralism. Of particular interest to many will be his discussion of place-names containing the *nemeton*-related word, with the original meaning 'sacred grove, sanctuary'. Here he develops an idea put forward by W.J. Watson in 1926 (Watson 1926, 244–50), adding several more *nemeton*-place-names to Watson's list, and for the first time producing a distribution map of the element.

His thoughts on the interpretation of place-names with the Scottish Gaelic element *monadh* are of particular importance, given its ubiquity in Scottish toponymy (see below pp. 62, 65–6). I offer the following as a very concrete example of how toponymics and history can work together, and as the result of a train of thought inspired by hearing the paper which forms the basis of Chapter 2.

It concerns the interpretation of one of the earliest recorded Scottish place-names, Kilrymont, which occurs in the Annals of Ulster as early as 747 (as *Cinrigh monai*).[1] This is the place which we know today as 'St Andrews' in Fife, the most important church centre in Scotland from the tenth century until the Reformation in 1560, and the site of Scotland's first university, founded in 1411. Kilrymont, earlier *Kinrymont, is a Scottish Gaelic name meaning 'at the end of the king's *monadh*'. It has led to much speculation as to what the *monadh* might be, and suggestions have ranged from 'burial mound' to 'hill', with reference to Scoonie Hill, the long ridge lying south of the town. Neither of these makes good sense – the first for semantic reasons, the second for topographic ones. However, if we follow Barrow's suggestion that *monadh* in eastern Scottish place-names frequently signifies 'rough grazing' or *muir* (a Scots word meaning 'rough grazing or permanent pasture'), then the meaning of *Kinrymont would be 'end of the king's *muir*'; with reference to a tract of upland grazing which runs several miles south-east from Scoonie Hill towards Crail. In fact, as soon as we leave the lands with which the early church at Kilrymont was endowed, we find ourselves in Kingsmuir, in upland Crail, a name which we can now understand as the exact Scots equivalent of *Righ Monai (rígh mhonadh)*. From all this, a geo-political picture begins dimly to emerge for this part of early medieval Fife: a large tract of land, chiefly valuable grazing, was taken out of royal demesne and given to the church at Kilrymont, no doubt a royal foundation. However, land adjacent to this land to the south-east (i.e. Crailshire) remained in royal hands long enough for the designation *rígh mhonadh* to be translated into Scots as *Kingsmuir*, probably in the thirteenth century. This interpretation of the place-name evidence in fact accords extremely well with the sparse historical records that have survived for this period.[2]

Introduction

Historical Geography

Margaret Gelling has made a major contribution to English toponymics on several different fronts, not least in the field of landscape history and settlement.[3] In **Chapter 3** she develops and refines ideas already advanced in her *Place-Names in the Landscape* (Gelling 1984). Her basic thesis is that the topographic vocabulary of the early Anglo-Saxon settlers – the terms they used to describe landscape features – was highly nuanced and exact, and conveyed information not only about height and shape, for example, but also about potential for settlement or exploitation. The research which she and her colleague Anne Cole have carried out, and continue to carry out, in England should be an inspiration and a guiding light to everyone working in toponymics, since there is no suggestion that the Anglo-Saxons were unique in this precise usage of topographic terminology. Those with an interest in the link between topography and language – and is this not the essence of place-name studies? – should be out in the field 'doing a Gelling and Cole' on the landscape, i.e. examining and comparing features which contain the same generic element over wide areas of the country. In Scottish Gaelic there is much scope for such work with a variety of topographic generic elements: *druim* and *gasg*, for example, both of which can be translated into English as 'ridge', but which may well have different applications. W.J. Watson, in the single most important book ever written about Scottish place-names – *The Celtic Place-Names of Scotland* – does suggest a particular meaning for *gasg*, but not every *gasg* corresponds to Watson's definition 'tail-like point of land running out from a plateau' (Watson 1926, 500): take for example the Gask Ridge in Strathearn, which runs almost the whole length of the strath (*c.*15 kilometres), but at neither its north-east or south-west end does it run out of a plateau.

Other topographic words which would repay such a study are Gaelic and Gaelic derived Scots words for 'pass, hollow in hill, etc.', such as *glac* (borrowed into Scots as *glack*), *slochd* (borrowed into Scots as *slock, slug,* etc.), *bealach* and *làirig;* the many Gaelic bog-words, such as, *bòg, eabar, gaoth, mòine* and *riasg;* also two Pictish bog-words: one, **gronn*, almost certainly borrowed into Scottish Gaelic; and **mig,* which seems not to have been borrowed. Such a study could also be usefully applied to fortifications and fortified sites, such as *dùn, longphort,* and *ràth,* and would have obvious relevance for the archaeologist (for which see also below next section). In fact, almost every element mentioned in Watson's important chapter entitled 'Some General Terms' (Watson 1926, 477–513) could be usefully subjected to the 'Gelling and Cole' approach.[4]

There can be no doubt that in Scotland, at least, there are many dialectal variations to be observed in the use of these mainly topographic terms. However, such variations, both synchronic and diachronic, will only be fully revealed and understood by the methods put forward in this chapter.

Archaeology

The usefulness and use of place-names for the archaeologist has long been recognised. In an English context, there is an excellent discussion of place-names and archaeology in Cameron 1963, where mention is made of the exciting discovery of a Roman tessellated pavement at a place called Fawler in Oxfordshire: Fawler is from the Anglo-Saxon meaning 'multi-coloured pavement'![15] Such names are like descriptive labels attached to sites of actual or potential archaeological significance. In 1995 the 'Graham Webster Laurels' prize in the British Archaeological Awards was won by the Herefordshire Field-Name Survey, whose collection and publication of 125,367 field-names from the 260 parishes of Herefordshire have already facilitated a variety of archaeological discoveries ranging from an Anglo-Saxon palace to early iron working, and will undoubtedly contribute to many more in the years to come (see *Current Archaeology*, 145 (1995), 11–15).

In Scotland, there are many elements that can and do act as archaeological indicators: these range from Scottish Gaelic words denoting various kinds of fortified and/or high status dwellings, such as *ràth, dùn,* and *lios,* to Scots words relating to such important former industries as cloth-waulking (fulling) (*waulkmiln*) and lime production (*limekiln*). Prehistoric monuments such as standing stones and burial cairns are clearly indicated by place-name elements such as the now obsolete Scottish Gaelic *coirthe* 'standing stone',[6] *càrn* 'cairn, burial mound',[7] or *cair '(Roman) fort'[8] and we can assume with reasonable certainty that the occurrence of these and similar elements in a place-name means such monuments once existed at that place, even if there is no obvious trace visible in the modern landscape.

The archaeological information being made available through aerial photography is particularly relevant in this context. Excavations cannot keep pace with this rapidly increasing body of material, and archaeologists will have to turn more and more to place-names as an interpretative tool. To take just one example, the extensive enclosures which have been aerially photographed at Ramornie Mains in Kettle parish FIF, are no doubt traces of the *ràth* contained in the place-name, as are the less clear remains of an enclosure at Rameldry, another *ràth*-name in the same parish (see Taylor 1995a, 278–9). Such close correlation between these elusive remains and a place-name element meaning 'high-status dwelling enclosed by earthen rampart(s)' is a clear signal to the archaeological community to look more closely at other *ràth*-place-names (Taylor 1995a, 71–3, 466–7), and the same can be said for all the elements mentioned above, as well as for many others.

So close and so important are the links between place-names and archaeology that two chapters have been devoted to the subject: in the first, **Chapter 4**, Terry James surveys the impressive results of the close collaboration between place-names studies and archaeology in South Wales: such collaboration can be place-name led, such as the discovery of the hitherto unknown archaeological feature indicated by the name Castell Draenog (see p. 108); or it can, in conjunction with an observable feature, help to interpret that feature, such as the *sarn* ('embankment') place-names along the line of a recently discovered Roman road west of Carmarthen (see pp. 115–16).

James' article also reminds us of the importance of medieval administrative units in the interpretation of place-names, as well as that of the underlying linguistic and political history.

In the most northerly inhabited island in the British Isles – Unst in Shetland – Steffen Stummann Hansen and Doreen Waugh are just beginning a joint archaeological and place-name venture which is fully described in **Chapter 5**. Again the symbiosis between archaeology and place-name studies is clearly brought out: not only can place-names lead archaeologists to potentially rich sites, and help in the interpretation of that site, but also archaeology can help define the way in which certain place-name elements were being used, as well as help date the antiquity of the use of these elements.

Literature

In **Chapter 6** Professor Donald Meek explores the complex interactive relationship between literature and landscape, and its effect on place-nomenclature, by means of the Gaelic ballads of Scotland and Ireland. Within this context he looks at how place-names can give rise to a new narrative – the process known as *dindshenchas* (literally 'the lore of noble places') – as well as how place-names can fix an already known narrative in a particular landscape, sometimes changing the details to fit certain local names or features. And once a narrative has taken root in a particular locality, it can then generate new place-names, or alter existing ones.

For example, in his discussion of Beann Ghulbainn/Ben Gullipen names, not only do we see a precise correlation between the shape of a mountain and its descriptive name, the kind of topographic precision which Margaret

Gelling has already led us to expect in Chapter 2, but also this precise description, meaning 'snouty mountain', has come to fix a well-known story in its vicinity (see pp. 153–8).

Thus Chapter 6 addresses a place-name phenomenon which tends to get somewhat neglected in standard place-name works, but which has general application to all place-names in all cultures. In a less literary context, the phenomenon is usually known as 'folk-etymology', although the above-mentioned *dindshenchas* is being used more and more to describe it, at least in Ireland and Scotland. At its most basic, it shares with the serious 'scientific' place-name scholar the same impulse: the desire to explain and understand the inexplicable, to make transparent the lexically opaque. It leaves no stone unturned in its quest to give a satisfactory meaning to otherwise incomprehensible or lexically opaque place-names: apart from the literary context, described with such clarity by Meek in Chapter 6, it draws on history (both real and pseudo), local lore, and local characters. Also, there is hardly a place-name, in Scotland at least, which has escaped its clutches. Even where a word is lexically transparent, or where an adequate explanation offers itself to anyone with the most rudimentary knowledge of Scottish place-names, there is the further impulse of romanticising what is perceived to be an otherwise somewhat mundane place-name, often by association with a well-known people, event or figure in history. Mary Queen of Scots looms especially large in a Scottish context, responsible, we are assured, for such names as Blantyre (where her horse Blan became tired during her escape from the battle of Langside), Corstorphin, where she lost her *croix d'or fin*, her cross of fine gold, or Beauly, which she declared to be a *beau lieu*. In this last case, the etymology is in fact correct; unfortunately it is first recorded c.300 years before Mary's appearance on the scene.

Queen Mary vies with the Vikings as chief generator of Scottish place-names, especially in the Lowlands. In Fife there is an example of how a desire to associate a place-name with these people has led to an orthographic controversy which is still very much alive: the village of Dunshelt (historically correct spelling) by Auchtermuchty is spelt by many villagers, and on some road-signs, 'Dunshalt', in the quaint belief that the name means 'Danes' Halt'.[9]

A most entertaining book could be devoted to the folk-etymologies of place-names, reflecting cultural and religious concerns and allegiances of all kinds. Their inventiveness knows no bounds, as in one of my favourite examples, the derivation of the Fife parish-name 'Ballingry'. According to the Rev. Thomas Scott, minister of the parish in the late eighteenth century, 'it is a compound of the Gaelic word *Bal*, which is a village, and *inri*, being the initials found on those crosses erected often in the fields, in honour of Christianity, on which were inscribed J.N.R.J. *Jesus Nazarenus Rex Judicorum,* Jesus of Nazareth King of the Jews' (*OSA Fife*, 74).

Environmental Politics

Chapter 7 by Heather James of the Dyfed Archaeological Trust concerns an aspect of place-names which, although not part of the February conference, was felt to be a most appropriate conclusion to a volume devoted to the uses of place-names. It could not in fact have been part of the conference, as the conclusions of the Public Inquiry which form the crux of the chapter were not made public until November 1996. In it we see place-name evidence being used in the cut and thrust of a Public Inquiry, and how the skillful mustering of a well maintained place-name data-base can make a decisive contribution to the outcome of an environmental inquiry.

The Inspector's statement that 'merely to record place-names that used to exist can serve little purpose without the preservation *in situ* of the landscape to which they refer' (see below p. 177) will of course, as a generality, not be accepted by toponymists. However, it is extremely timeous and appropriate in the case of the disputed land in question – part of the parish of Llandybïe in Carmarthenshire – containing as it does one of the oldest recorded place-names in Wales. There is over time an inevitable loss of landscape, not simply through the drastic changes wrought by opencast mining, but also through massive urban and industrial development, especially in the UK in the past 200 years. In such drastically altered environments the recording and analysis of place-names can be a vital key in the reconstruction of lost landscapes.

However, the alarming expansion in the opencast mining industry of coal and other minerals, especially in the former deep-mining areas of southern Scotland, South Wales and northern and midland England, poses a threat to an environment already under severe pressures from many other directions; and the Inspector's conclusions in the Llandybïe Inquiry should be a guiding light wherever such applications involve a threat to historically significant landscapes, and the model tactics of the Dyfed Archaeological Trust an inspiration to all who care about the loss of our environmental heritage.

This final chapter of the book reminds us also that place-names are political, and there is much more that could be said in this respect, especially in the context of Cornwall, Northern Ireland, the Republic of Ireland, Scotland and Wales, where non-English languages try with more or less success to assert themselves against the overwhelming influence of the

English language, both historical and contemporary. The recent heated debate over Gaelic road-signs in the Gàidhealtachd of Scotland is a case in point, and it is only a matter of time before such an issue will have to be faced in a Scots-speaking context in the Scottish Lowlands.

A place-name has many uses, but, as is stressed over and over again in all the following chapters, both explicitly and implicitly, before it can be used effectively for whatever purpose, it cannot simply be plucked from a modern Ordnance Survey map or a gazetteer: a great deal of work has to go into a place-name before it is of use to any of the disciplines represented in this book. The one discipline which is not directly represented here is that of toponymics *per se*, which has its own methodology and academic rigour, and which underpins all other place-name related work. It is not the purpose of this book to look in any detail at the workings of this discipline: they have been well described elsewhere: for example in the opening chapter of W.F.H. Nicolaisen's *Scottish Place-Names* (Nicolaisen 1976) and the introduction to O.J. Padel's *Cornish Place-Names* (Padel 1988). These form excellent introductions to the problems and methods of place-name study, each in its own way transcending its geographical limitations. However, it is on the success or failure of this discipline that all other interdisciplinary uses of place-names hinge.

On the evening of the above-mentioned 'Uses of Place-Names' conference held in St Andrews in February 1996, the inaugural meeting of the Scottish Place-Name Society was convened, also in St Andrews, and attended by many of those who had been at the conference. One of the aims of this Society is to provide the academic world, both amateur and professional, as well as the interested general public, with a comprehensive place-name data-base which is properly researched and analysed. It is envisaged that this data-base would be both electronic and paper. In its electronic form it will be constructed in such a way that it can be fully exploited and satisfactorily interrogated by members of all disciplines, as well as be accessible to the casual enquirer. In its paper form, it will be modelled on such successful place-name series as that of the English Place-Name Society English County volumes, or the Northern Ireland Place-Name Project's *Place-Names of Northern Ireland*.

England began the systematic collection and analysis of its place-names with the foundation of the English Place-Name Society in 1923, and most of its counties now have a volume or set of volumes devoted entirely to their place-names. On the basis of this valuable resource, an electronic data-base of English place-name elements is being compiled at the University of

Introduction

Nottingham funded chiefly by the Leverhulme Trust. It could be said that Scotland in comparison has hardly begun its work. This is not to say that there are no good local, and national, place-name studies, but they are piecemeal and of very uneven quality, and leave large areas of the country virtually untouched by any kind of place-name analysis. The demands which are, quite rightly, now being put upon place-names, and the great potential which they can offer to all kinds of Scottish cultural, historical, linguistic and archaeological studies and activities, as testified to by the following chapters, can only be met by the systematic response proposed by the Scottish Place-Name Society, which deserves the fullest support on a national level.

Simon Taylor
St Andrews Scottish Studies Institute
University of St Andrews
September 1997

Abbreviations

ES *Early Sources of Scottish History 500 to 1286*, ed. A.O. Anderson (Edinburgh, 1922). (Reprinted with bibliographical supplement and corrections by M.O. Anderson, Stamford, 1990.)
OSA Fife *Statistical Account of Scotland 1791–99: Fife*, (vol. x), ed. J. Sinclair, reprinted with new introduction by R.G. Cant, Edinburgh, 1978.
PSAS *Proceedings of the Society of Antiquaries of Scotland* (1851–).

Bibliography

Barrow, G.W.S., *Kingdom of the Scots* (London, 1973).
Barrow, G.W.S., *The Anglo-Norman Era* (Oxford, 1980).
Barrow, G.W.S., 'Popular Courts in Early Medieval Scotland: Some Suggested Place-Name Evidence', *Scottish Studies* (1981), vol. 25, 1–23. [Also in Barrow, G.W.S., *Scotland and its Neighbours in the Middle Ages* (London, 1992), 217–45).]
Barrow, G.W.S., 'The Childhood of Scottish Christianity: a Note on Some Place-Name Evidence', *Scottish Studies* (1983), vol. 27, 1–15.
Barrow, G.W.S., 'Land Routes: The Medieval Evidence', in *Loads and Roads in Scotland and Beyond*, ed. A. Fenton and G. Stell (Edinburgh, 1984), 49–66. [Also in Barrow, G.W.S., *Scotland and its Neighbours in the Middle Ages* (London, 1992), 201–16, entitled simply 'Land Routes'.]
Barrow, G.W.S., 'The Lost Gàidhealtachd', in *Gaelic and Scotland*, ed. W. Gillies (Edinburgh, 1989), 67–88. [Also in Barrow, G.W.S., *Scotland and its Neighbours in the Middle Ages* (London, 1992), 105–26.]
Broun, D. and Taylor, S., *The Foundation Legends of St Andrews in Scotland* (provisional title), (forthcoming).
Cameron, K., *English Place-Names*, 2nd edn (London, 1963).
Cameron, K., 'Stenton and Place-Names', in *Stenton's Anglo-Saxon England fifty years on*, ed. D. Matthew *et al.* (Reading, 1994), 31–49.

Gelling, M., *Place-Names in the Landscape* (London, 1984).
Jones, B.L., 'Why *Bangor?*', *Ainm* (Bulletin of the Ulster Place-Name Society), (1993), vol. v, 59–65.
Nicolaisen, W.F.H., *Scottish Place-Names* (London, 1976; paperback edn 1986).
Nicolaisen, W.F.H., 'Gaelic and Scots 1300–1600: Some Place-name Evidence', in *Gaelic and Scots in Harmony*, ed. D.S. Thomson (Glasgow, 1988), 20–35.
Padel, O.J., *Cornish Place-Names* (Penzance, 1988).
Taylor, S., 'Some Early Scottish Place-Names and Queen Margaret', *Scottish Language* (1994), vol. 13, 1–17.
Taylor, S., 'Settlement-Names in Fife' (unpublished Ph.D. thesis, University of Edinburgh, 1995a).
Taylor, S., 'The Scandinavians in Fife: the Onomastic Evidence', in *Scandinavian Settlement Studies*, ed. B.E. Crawford (London and New York, 1995b), 141–67.
Watson, W.J., *The History of the Celtic Place-Names of Scotland* (Edinburgh, 1926; paperback edn 1993).

Notes

1 See *ES* i, 238, where the forms from all the relevant annals are usefully brought together.
2 For an up-to-date summary of all the material relating to the establishment of the church at St Andrews, see Broun & Taylor, forthcoming.
3 In K. Cameron's extremely useful review of English place-name studies over the last 50 years, (Cameron 1994) Margaret Gelling's name appears on almost every page.
4 For example, only the 'Gelling and Cole' approach will satisfactorily decide the interpretation of the place-name 'Banchory', since opinion is divided as to whether this place-name originally describes a topographical feature or an early religious settlement. See Watson 1926, 480–2 and Jones 1993; see also below p. 113.
5 Cameron 1963, 111–12. Unfortunately the chapter 'Place-Names and Archaeology' has been omitted from the new (1996) edition of this book.
6 For example in the various places called Pitcorthie and Pitforthie in Fife and Angus; Milnathort in Kinross-shire; or the many Aquhorthies, etc. in Aberdeenshire.
7 For example, two places called Pitcairn in Fife (in Auchterderran and Leslie); in the former the large bronze age burial cairn is still in existence; in the latter the cairn is now gone, but was excavated prior to its disappearance (see *PSAS* vol. 109, 361–6).
8 For a discussion of the much neglected **cair*-element in place-names north of the Forth, see Taylor 1994, 8–10; also Taylor, 1995a, 67–71, 443–5.
9 For more discussion on this name see Taylor, 1995b, 141–2; see also *Scottish Place-name News*, 1 (the newsletter of the Scottish Place-Name Society), 8–9.

1

Place-names as a Resource for the Historical Linguist[1]

Roibeard Ó Maolalaigh

Place-names bear relevance to a wide range of disciplines, including not least linguistics. They contain the fundamental building blocks of language and as such have intrinsic value as linguistic items even if the linguistic content of such items is somewhat limited. My aim is not, however, to discuss place-names from a general linguistic point of view. Place-names are of considerable importance for tracing the history of Gaelic, Scots and English in Scotland. Despite this fact, the Scottish place-name evidence remains a relatively under-exploited resource. To date, much work has focused on the geographical distribution of particular place-name elements with a view to establishing the boundaries and frontiers of the various cultural and social groups resident in Scotland during the historical period and even beyond it. The present chapter is an attempt to illustrate how place-names may be exploited by the historical linguist, with particular reference to Gaelic but also, where relevant, to Scots. We will see, from a purely linguistic standpoint, how place-names can provide a window on the past in a way which other historical sources frequently do not. A brief consideration of the historical socio-linguistic situation of Gaelic in Scotland should highlight the importance of Scottish place-names in the present context.

Gaelic was introduced into Scotland some time before the sixth century AD by Irish settlers. The traditional view has generally been that the forms of Gaelic as spoken in Ireland and Scotland remained substantially the same until the formative period of Gaelic dialects in the thirteenth and subsequent centuries (O'Rahilly 1932; Jackson 1951). When the historical linguist turns to the available literary sources of this period in order to observe the development of Gaelic, he is confronted with a formidable barrier. The thirteenth century coincides with the beginning of the second major period of standardisation of the Gaelic language, which is usually referred to as the

Classical Irish or Early Modern Irish period. During this period there was in place a standard literary language which was created and mainly used for the composition and transmission of official court poetry. The standard language, usually referred to as Classical Irish, was used in Ireland until the seventeenth century and in Scotland until a century or so later. This codified high-register literary language coupled with the inherited conservatism of the Gaelic written tradition served to conceal traces of the emerging vernacular language(s) from the written record. The written language of the Classical period, propagated by the Bardic schools, was in many ways a highly conservative one and for more than four centuries was fairly rigidly maintained and adhered to by the learned orders of both Ireland and Scotland. The conservatism of this standard language can be illustrated by the pronunciation of the dental fricatives *th* and *dh*. These sounds had lost their dental articulation in vernacular speech by the end of the thirteenth century in Ireland (O'Rahilly 1930, 1932) and probably also in Scotland (Jackson 1972, 55, 63; Taylor 1995, 42–3). However, it is clear from the requirements of alliteration and rhyme in the strict syllabic verse of the Classical period (1200–1650) that these sounds probably continued to be pronounced as dentals in certain high registers during the following four or five centuries. A typical alliterating pair in Classical Irish verse might be *thoil* /θol'/ ~ *tobar* /tobər/ and a typical rhyming pair would be *crích* /k'r'i:χ'/: *síth* /s'i:θ'/, where *th* is given a dental value similar to English *th* in *thorn*.

Although the Classical language offers little insight into the vernacular Gaelic language(s) as spoken during the Classical period, it does tell us a considerable amount about developments in certain varieties of Irish (and in some cases by implication also in Scottish Gaelic) during the Middle Irish period, i.e. 900–1200. Classical Irish, an artificially planned language created towards the end of the twelfth century, could be described as a judicious blend of older inherited linguistic forms and newer or more innovative forms attested occasionally in the Middle Irish period (900–1200) (Ó Cuív 1973, 1980; McManus 1994). In Classical Irish, old and new forms are often found side by side, and their occurrence is frequently dependent upon the necessities of rhyme and/or alliteration. Innovative forms reflect the major linguistic developments which had occurred in the Irish language since the Old Irish period. One could cite as examples the rise of 'new' analytic verbal forms (e.g. *molaidh siad* 'they praise') alongside older synthetic verbal forms (e.g. *molaid* 'they praise') (Ó Cuív 1980, 24; McManus 1994, 419–20); or the spread of dental plural allomorphs *-(e)adha*, e.g. *cridheadha* 'hearts' alongside older *cridhe* 'hearts' (Greene 1974). You will note that I refer to linguistic developments in the Irish language, not in the Gaelic language for it is the contention of the present author that the Classical norm was created on Irish soil by Irish poets who drew solely on the resources of the language as it was spoken and written in Ireland. In short, there are no innovative features of Classical Irish which cannot be explained in purely Irish linguistic terms.

The methodology of comparative reconstruction has in recent years produced a convincing body of evidence, modifying Jackson (1951), for the early divergence of Irish and Scottish Gaelic in some important respects and a clearer picture of the structure of pre-twelfth century Scottish Gaelic is gradually emerging.[2] There is a significant number of distinguishing features between Irish and Scottish Gaelic whose origins must be quite early, many of them certainly much earlier than the twelfth century when the Classical language was supposedly created. Such features include the plural allomorph {*an*} (Greene 1974; Ó Buachalla 1988), the pronominal system (Ó Sé 1996), the verbal system (Greene 1972; Macaulay 1975) and perhaps also the system of eclipsis (on which see below) (Ó Maolalaigh 1995/96). However, none of the Scottish features which we would now associate with pre-twelfth century Scottish Gaelic are represented in the Classical Irish language. The absence of these Scotticisms in the Classical language may, in the past, have led some scholars to believe that these features had not yet evolved by the twelfth century. However, there are other more plausible explanations for the absence of Scotticisms in the canon of Classical Irish. If we accept that the Classical language is an Irish creation, then the absence of Scotticisms must be understood as being due to the tendency of the creators of linguistic norms to build on 'central' features at the expense of 'peripheral' features. It follows that the Irish language planners of the twelfth and subsequent centuries perceived the Scottish variety of Gaelic as being peripheral. The lack of overt Scottish features in the Classical language no doubt contributed to and consolidated the poor-sister status of Scottish Gaelic within the Gaelic world from the thirteenth century onwards. The fact that Scottish poets wrote and composed verse through the medium of Classical Irish should not be taken as evidence, as has so often been the case, of the non-divergence of Irish and Scottish Gaelic during this period. The bottom line is that Scottish poets were educated in a language which was firmly based in twelfth century or earlier Ireland linguistically.[3]

However much we may admire the resourcefulness of the original language planners in the twelfth and subsequent centuries, their standard language had the result of concealing the emergence of the modern Gaelic languages. In particular, and more importantly for our purposes, the standard language served to obfuscate the existence of an emerging Scottish Gaelic vernacular. It is not until the sixteenth and seventeenth centuries, with relatively few exceptions before then, that we begin to get a glimpse of the modern vernacular languages in both Ireland and Scotland.[4] It is against the briefly sketched background presented above, namely the conservatism of the Gaelic written tradition and its exclusion of Scottish Gaelic features, that the importance to the Gaelic historical linguist of ancillary evidence such as place-names should be understood.

We have already noted that place-names can provide a window on the past in a way which conventional literary sources frequently cannot. Place-names may contain fossils or relics which reflect earlier stages or developments in a particular language, whether in terms of grammar, phonology ('the sounds of language') or semantics ('the meaning of words'). Such fossils are formed in various ways. Words may develop different meanings in the normal lexicon or perhaps even become obsolete, thus leaving fossils embedded in the onomastic lexicon. Fossils are perhaps more commonly formed in language-contact situations, that is to say, when place-names are 'borrowed' from one language into another and are thus no longer subject to the linguistic laws of the 'donor' language. As long as due consideration is given to the 'new' linguistic environment, the linguistic structure of 'borrowed' place-name elements can offer valuable insights into earlier stages of the 'donor' language.

The history of Gaelic and Scots provides a valuable example of the language-contact phenomenon which results in the process of fossilisation just described. Early contact between the two languages has left a vast number of fossilised Gaelic words in the toponymy of Scots- and English-speaking Scotland. Furthermore, these fossils can in many cases be traced through time in the historical sources, in some cases as early as the twelfth century. The corpus of Gaelic place-names, both historical and current, surviving in Scots and English provides the historian of Scottish Gaelic with a potentially rich resource for investigation during a period whose conventional literary sources are, as has already been noted, characterised by their linguistic conservativeness. Indeed it is only when ancillary sources such as these have been meticulously scrutinised and properly documented that we can hope to establish early dialect divisions – impossible at present – and thus gain a fuller insight into the early history of Scottish Gaelic.

It would of course be unwise to exaggerate the usefulness of place-names to the historical linguist. The linguistic content of place-name evidence is after all severely limited. The basic structure of the majority of place-names is that of the simple noun phrase. Furthermore the use of place-names for historical linguistic purposes, particularly when attested outside their natural linguistic environment, is fraught with many difficulties. The interpretation of historical forms, for instance, runs into problems of scribal practices, not to mention scribal and editorial errors. Can we be certain what a scribe intended to signify with a particular symbol or set of symbols, particularly when attempting to represent a 'foreign' sound or word? Nicolaisen (1990, 23) warns us about the 'fundamentally onomastic nature' of place-names:

> When linguists or students in other adjacent disciplines look to names, especially place names, for information which words cannot provide, their legitimate desire to exploit toponymic evidence has to be matched by their sobering recognition of the fundamentally onomastic nature of the evidence whose help they seek. Otherwise, faulty or inflated expectations are bound to lead to frustration and disappointment. There are certain things which names do well and others that names do badly, like absolute dating, for example.

With these salutary words in mind, I would, at this stage, like to anticipate some of the pros and cons which may be associated with the use of place-name evidence for historical linguistic purposes.

Advantages

i. Place-names provide valuable information (albeit to varying degrees as we shall see) on phonology, noun morphology (inflectional and derivational), and the interface between these, i.e. morpho-phonology. In simple terms this means pronunciation, basic noun phrase formation and inflection, and initial mutations such as lenition and eclipsis (on which see below pp. 22–30).

ii. The syntax of place-name morphology can provide valuable substratum information. In Scotland, the basic form of generic followed by the stress-bearing specific is a good (though not necessary) tell-tale sign of an original or underlying Celtic/Gaelic onomasticon.

iii. Fossilised forms frequently survive in place-names and thus provide valuable glimpses of earlier forms of language, now obsolete in the donor language, e.g. neuter gender (discussed below p. 19).

iv. Datable historical forms can provide a *terminus post* (or *ante*) *quem* for many sound changes.

v. The modern form of borrowed place-name elements can tell us when contact first occurred between the donor and receiver language.[5]

vi. Place-names can contain words which are otherwise unattested, or which are used with a meaning different from that found in other sources.

Disadvantages

i. Although place-name evidence may provide a *terminus post quem* for a particular linguistic development, absolute dating is of course impossible.
ii. Historical spellings in our sources are often ambiguous.
iii. The conservative nature of the spelling tradition of place-names is a serious problem.
iv. It has been suggested that place-names may not be subject to the same linguistic rules as ordinary lexical items (Nicolaisen 1990, 23–4).[6]

Even though the linguistic content of place-names is restricted from a general linguistic point of view, valuable information can, nevertheless, be culled from them. Most, if not all, place-names can be classified as noun phrases, the majority of which contain a noun or a string of nouns with identifiers (e.g. the article) and describers (e.g. adjectives). Some place-names do of course contain prepositional phrases but on the whole such instances are quite rare. An amusing and instructive example of such is to be found in the Irish names Tanderagee (ARM), Tonragee (MYO), Tonlegee (CAV), etc. where they usually refer to the administrative unit of a townland (Ó Maolfabhail 1982, 366). There are approximately 45 place-names with this name in Ireland according to Ó Maolfabhail (1982), the majority of which occur in Connacht, north Leinster and in Ulster. The place-name is also found in south-west Scotland, e.g. Tandragee (WIG) (Maxwell 1930, 256), Thundergay (in Gaelic *Torragaoth*)[7] (Arran ARG) (Holmer 1957, 62).[8] These place-names are normally derived from the phrase *Tóin re Gaoith* which is usually interpreted as 'backside to the wind'. This particular example is a good instance of the way in which the original meaning of a place-name can come to be transformed by popular folk etymology. It has been argued that 'podex ad ventum' may not have been the original meaning of this set of names. Ó Maolfabhail (1982, 371) discusses some of the technical, originally metaphorical, uses of *tóin* in Irish place-names. Noting that the majority of places bearing the name occur in the vicinity of water, he argues that the original meaning of *gaoth* must have been 'water, inlet, etc.' rather than 'wind'. When *gaoth* ceased to mean 'water' in the Irish lexicon, the *gaoth* element naturally came to be associated with the homophone *gaoth* meaning 'wind'.[9] That *gaoth* may have originally referred to 'wind' in some of the Tanderagee–Tonlegee names cannot be discounted. Flanagan (1994, 90) who takes a rather literal meaning from the name notes: 'it [*gaoth* "wind"] appears, almost as a warning, in Tanderagee, Co. Armagh, which is *Tóin re Gaoith*, "Backside to the wind", as if to suggest that is the right stance in the area'!

Leaving aside the semantic problems inherent in the name, we may note that Tanderagee–Tonlegee place-names provide the historical linguist with some potentially useful information on one aspect of the history of Irish. This set of place-names may be classified into two groups, those which contain -*r*- and those which contain -*l*-. The -*r*- forms provide an interesting fossil from the linguist's point of view since they contain the preposition *re* 'against, before' (deriving from the prepositions *fri* and *ré*) which is now obsolete in all Irish dialects.[10] The preposition *re* has been replaced by the preposition *le* meaning 'with' which we see in the -*l*- forms of the place-names under discussion. Most Scottish Gaelic dialects on the other hand have preserved both *ri* and *le* in their original meanings of 'against, to' and 'with' respectively, although there is evidence of their having partially merged in some dialects.

A survey of the geographical distribution of -*r*- and -*l*- forms is revealing.[11] Forms with -*r*- are restricted to north west Connacht (Mayo, Roscommon, Sligo),[12] Ulster (Donegal, Antrim, Tyrone, Armagh, Fermanagh, Monaghan, Cavan) and north Leinster (Meath, West Meath). From a synchronic point of view there is clearly a bias for -*r*- forms in the northern parts of Ireland. However, it is interesting to note that many of the current -*l*- forms show variation between -*r*- and -*l*- forms in the historical record, e.g. Tonlegee (WMH) occurs as *Toneregihy* (1581), *Tonlegeihe* (1584), *Tonregee* (1634), *Tonlegihie* (1659), *Tonlegee* (1663) (Ó Maolfabhail 1982, 377).[13] The current forms and their geographical distribution when considered alongside their historical forms, suggests: (a) the original form of the name contained the preposition *re* rather than *le*; (b) the replacement of *re* by *le* in this group of place-names may have originated in the southern parts of Ireland; (c) the change *re* to *le* in this place-name may be a relatively late one which had begun by the sixteenth century at the latest. Tonlegee (WMH) (*Tonregihy* [1581]) occurs with -*l*- first in 1584 and thus provides a tentative *terminus ante quem* for the change.[14]

Although it would be imprudent to push the Tandragee, Tonlegee, etc. evidence too far, it nevertheless provides a tentative framework against which the historical development of the preposition *re* in Irish may be discussed and perhaps eventually understood. Our discussion of the place-name evidence suggests that the change *re* to *le* may have originated in southern Irish dialects some time before or during the sixteenth century and spread northwards.[15]

The linguistic content of place-names
Our discussion of place-names as a resource for the historical linguist will cover the following topics: (1) fossilised forms of language retained in place-names: the cases of (A) the neuter gender in Gaelic and (B) old forms of the definite article in Gaelic; (2) initial mutations in Gaelic: (A) lenition and (B) eclipsis; (3) *-in* ~ *-ie* variation in Scottish place-names; (4) Gaelic *-ach* in Scottish place-names.

1. Fossils
(A) Neuter gender
In Old Irish there were three genders: masculine, feminine and neuter. Neuter nouns eclipsed (i.e. changed the value of) the initial consonant of a following word and prefixed *n-* to a following vowel. There is evidence for the disappearance of the neuter category during the Middle Irish period (i.e. 900–1200) (Breatnach 1994, §4.13, §4.14, §5.4). Fossilised remnants of the neuter survive more commonly in place-names than in the modern languages[16] and tend to occur most commonly when a noun contains an initial vowel or *c-*. Instances of *n* occurring before vowels are problematical as they may in some cases derive from genitive forms of the article:

Scotland
Invernauld	*Inbhear **n**Allt* (SUT)
Cumbernauld	*Combor **n**Allt* (DNB)[17]
Loch Nell	*Loch **n**Eala* (ARG)[18]
Loch Neldricken	*Loch **n**Eileireag* (KCB)[19]
Glen Noe	*Gleann **n**Abha* (ARG)
Drumgoudrom	*Druim **g**Colldroma* (ABD)[20]
Dundurcus	*Dùn **d**Turcais* (BNF)[21]

Other possibilities might include:
?Lochnaw	*Loch **n**Abha* (WIG)
?Lochnawean (Hill)	*Loch **n**Abhainn* (KCD)
?Dundee	*Dùn **d**Toí* ('of Tay') (ANG)

Ireland
Lough Neagh	*Loch **n**Eachach* (ANT)
Moynalty	*Magh **n**Ealta* (MTH)
Mullingar	*Muileann **g**Cearr* (WMH)
Slieve Gullion	*Sliabh **g**Cuillinn* (ARM)
Loughguile	*Loch **g**Caol* (ANT)
Moygashel	*Magh **g**Caisil* (TYE)

(B) Old forms of the article

The inherited paradigm of the definite article has been simplified in the modern Gaelic languages. In particular, the older forms *inda* (genitive plural) and *ind* (masculine genitive singular etc.) have disappeared and have been replaced by *na(n)* and *an* respectively.[22] Pokorny (1923) has shown how the older *inda* has been preserved in some place-names but has been reinterpreted as *an* (singular article) + *dá* (the numeral 'two') e.g. Cloondanagh (CLE) from *Cluain inda nEach* which originally meant 'meadow of the horses'. This has been reinterpreted as *Cluain an dá n-Each* 'meadow of the two horses' (Hogan 1910, 26). Similarly Rosdanean (FER) from *Ros inda nÉan* 'promontory of the birds' has come to mean 'promontory of the two birds' (Hogan 1910, 585). Some possible examples from Scotland include: Glendaruel (ARG) (*Gleann inda Ruadhail* 'glen of the red spots'[23]; Dundarave (ARG) (*Dùn inda ràmh* 'fort of the oars'; it is possible that *rave* may derive from *lamh* 'hand', cf. next example); Dundalav (INV) (*Dùn inda lamh* 'fort of the hands', although 'two hands' is as plausible); Dun-da-gu (Mull ARG) (*Dùn inda gaoithe* (singular) or *inda gaoth* (plural) 'fort of the wind(s)').

The older form *ind* (masculine genitive singular article) may have been preserved in a small number of Scottish place-names.[24] It has been claimed that *ind* appears in a twelfth-century recording of an Aberdeenshire place-name which is now obsolete. The form in question is *Petenderleyn* (which also occurs as *Petferlen, Petferlane*) for Gaelic *Pet ind fhir léighinn* 'the estate of the man of learning/lector' (Watson 1926, 267; Alexander 1952, 99).

Jackson (1972, 145) discusses the Aberdeenshire place-name *Lurgyndaspok*, now also obsolete, which occurs in a late fourteenth century source (1391). He conjectures that this represents Gaelic *Lorg an t-Easbag* ('of the bishop') with nominative *an t-Easbag* for expected genitive *an Easbaig*. This interpretation is also implicitly suggested in Alexander (1952, 328). Jackson's interpretation implies that the inherited nominal case system had been drastically reduced in fourteenth-century Aberdeenshire Gaelic. Indeed he goes on to say that 'the breakdown of the case-system, and the consequent use of the nominative as an all-purpose case, is a familiar feature in some modern Sc. G. dialects, above all in the eastern parts of the mainland; and its beginnings may be traced to the end of the fourteenth century if the place-name *Lurgyndaspok*... is reliable' (Jackson 1972, 145).[25] I would like to suggest that Jackson's derivation of *Lurgyndaspok* is in fact unreliable, the implication being that we cannot speak of the collapse of the case system in fourteenth-century Aberdeenshire Gaelic. The internal *-d-*

which occurs in *Lurgyndaspok* is more plausibly explained if we posit the presence of the older form of the article *ind* as has been suggested for *Petenderleyn*. If this is correct, this place-name would then derive from Gaelic *Lorg ind Easpaig*, with the expected genitive form *Easpaig*.

Jackson (in Barrow (1974, 38) appears to invoke the breakdown of the case system once again in a suggestion made to Professor Barrow in his explanation of a thirteenth-century Fife place-name which occurs variously as *Tarveht in dan* (1260), *Taruetandane* (c.1280), *Taruetadan'* (c.1290). This he apparently derived from 'Tarvit *an t'abha(i)nn* (*sic*) or (with *t* voiced to *d* after *an*) *an d'abha(i)nn* (*sic*), i.e. "Tarvit of (=by) the river"'. This derivation is objectionable on various grounds, the most important of which is that it implies the use of the masculine article *an t-* (*recte*) with the feminine noun *abha(inn)*, which so far as I am aware is nowhere attested in the recorded history of the word. Furthermore, in those modern Scottish Gaelic dialects where there is evidence of the collapse of the traditional case system, the generalised masculine article *an t-* is used usually only with historically masculine nouns, not with feminine nouns. See Dorian (1978, 93; 1981, 129–36).[26]

I would derive *Tarvitindan from Gaelic *Tarvet ind fháin* meaning 'Tarvit of the slope/low lying place' containing the older form of the article *ind* and *fán* meaning 'slope'.[27] Simon Taylor has suggested to me that this derivation appears to be supported by the fact that it is coupled with the place-name *Tarvitanaird ('Tarvit of the height') which occurs as *Taruetanard'* alongside *Taruetadan'* in the same source (c.1290) (Barrow *op. cit.*). The coupling of *fán* with *ard* is also attested in other place-names. Simon Taylor has kindly provided two other examples: Ballindean in Balmerino parish in north-east Fife (*Ballindan* 1232 *Balm. Lib.* no. 1; *Ballindan* 1230x38 *Balm. Lib.* no. 37; *Balnedan* 1234x41 *Balm. Lib.* no. 7 etc.), which contrasts with neighbouring Grange, formerly *Ballindard, now obsolete (*Balnedart* 1234x41 *Balm. Lib.* no. 7);[28] and Ballindean, Inchture parish, in the Carse of Gowrie PER, which occurs as *Balendan c.*1200 (*St. A. Lib.* 270). Another example may be Bellendean, Roberton ROX (*Bellenden* 1624 RMS). *Fán* and *ard* are also coupled together in an interesting Old Irish gloss contained in the ninth century Milan Glosses in a Latin commentary on the Psalms: *etir fán 7 ardd* (Ml 140a2), which is translated by the editors as 'both valley and height' (Stokes 1901, 474).[29] I am not aware of any examples of the pairing of *fán* and *ard* in Irish place-names.

2. Initial Mutations
(A) Lenition
Lenition is an initial mutation whereby certain consonants (with the exception of *f-*) are replaced by phonologically related fricative consonants (usually homorganic). Lenition is indicated in Gaelic orthography by inserting the letter *h* after the affected consonant e.g. *ch-, th-, ph-, gh-, dh-, bh-, mh-, sh-, fh-*. Lenited *f-*, i.e. *fh-* is silent. A significant characteristic of the development of the Gaelic languages has been an increase in the use of lenition in terms of the syntactic environments in which it may occur. In particular, lenition has spread in the modern languages to genitive forms of proper masculine nouns when preceded by nominative masculine nouns with a close syntactic relation to the genitive noun, e.g. *Baile Dhomhnaill* 'homestead of Donald' for earlier *Baile Domhnaill*.

The earliest datable examples that I am aware of which illustrate the innovation in question come from place-names. The two examples which I have, one from Ireland and the other from Scotland, in a remarkable coincidence, are dated to approximately the same year, 1157. These are *Caselanagan* for modern Castle Enigan (DWN) from Irish *Caiseal Fhlannagáin* (Toner 1992, 14) and *Balheluy* for modern Belhelvie (ABD) from Gaelic *Baile Shealbhaigh* (Watson 1926, 137; Alexander 1952, 18). It is not without significance that both examples contain instances of the two consonants in Gaelic which when lenited do not yield homorganic fricatives.[30] It should be noted, however, that the lenition in these examples may reflect locative forms where lenition would be expected.

In other cases lenition is not always consistently represented in anglicised place-names in Ireland or in anglicised or scotticised place-names in Scotland, where lenition would be expected from a historical point of view.[31] Where variation occurs in place-name sources between lenition and non-lenition, the current form in use today almost invariably reflects the form without lenition, e.g. Kilbroney (DWN) from Gaelic *Cill* (fem.) *Bhrónaighe* which appears in 1366 as *Killwronaygh* (thus reflecting lenition) but fourteenth to twentieth century place-name sources consistently show forms without lenition (Toner 1992, 132–5). Similarly, Ballybranigan (DWN) from Irish *Baile Uí Bhranagáin* (with lenition) appears with and without lenition in the historical record as the following representative examples illustrate: *Ballybranigan* (1588), *Ballyvranegane* (1625), *Ballyvranigan* (1662), *Ballybraniken* (1723) (Hughes 1992, 56–7). Other illustrative examples of the presence and absence of lenition include:

Scotland
Lenition shown:
Ben Vane	*Beinn Mheadhain* (DNB, Irving 1928, 19)
Knockvadie	*Cnoc an Mhadaidh* or *Cnoc Mhadadh* (DNB, Irving 1928, 41)
Balvaird	*Baile an Bhaird* or *Baile Bhard* (FIF, Taylor 1995, 45)
Balveny	*Baile Bheathain* (BNF, Watson 1926, 312)
Kilpheder	*Cill Pheadair* (INV)

Lenition not shown:
Kilpatrick	*Cill Phàdraig* (DNB)
Auchencairn[32]	*Achadh an Chairn* (KCB)
?Pitcruive	*Peit Chraoibhe* (PER)

Ireland
Lenition shown:
Tulrahan	*Tulach Shrutháin* (MYO)
Tullyvin	*Tulach Bhinn* (CVN)
Tullaherin	*Tulach Thirim* (KLK)

Lenition not shown:
Shankill	*SeanChill* (ANT, DWN)[33]
Ballincrea	*Baile an Chraoibh* (KLK)
Coolcullen	*Cúl an Chuilinn* (KLK)
Kilfeaghan	*Cill Fhéichín* (DWN)

(B) Eclipsis

Eclipsis is an initial mutation whereby initial consonants are replaced or 'eclipsed' by phonologically related phonemes. The consonants which are normally affected include the stops *c-, t-, p-, g-, d-, b-* and the voiceless labial fricative *f-*. Eclipsis prefixes *n-* to vowels. Eclipsis in Irish and Manx changes initial *c-, t-, p-, f-* to the corresponding homorganic voiced sounds *g-, d-, b-, v-* and changes initial *g-, d-, b-* to the corresponding homorganic nasal sounds *ng-, n-, m-*.[34] Eclipsis is quite different in Modern Scottish Gaelic where it is not represented in the orthography. While the realisation of eclipsis is uniform throughout Irish dialects, the situation is more complex, phonetically speaking, in Scottish Gaelic.[35] For the purposes of the present chapter we may simply note that eclipsis in Scottish Gaelic voices (albeit to varying degrees) the stops *c-, t-, p-* and, unlike Irish, does not fully nasalise the stops *g-, d-, b-*. We may also note that eclipsis in Gaelic occurs following the preposition *i* (Ir), *an* (ScG) 'in', and the genitive plural article *na* (Ir), *nan* (ScG).

It has until quite recently been accepted that the Irish system of eclipsis once existed in Scottish Gaelic and that the Scottish Gaelic system represents a later restructuring of the inherited Irish system. The traditional orthodoxy had little to do with solid linguistic facts and more to do with a received ideology with regard to the status of Scottish Gaelic in relation to Irish. Preconceptions are markedly clear from statements like the following: 'Most remarkable of all Scottish Gaelic *peculiarities* [my italics] is its treatment of the initial mutation ('eclipsis') of consonants.' (O'Rahilly 1932, 150.)

It has been argued in a recent article (Ó Maolalaigh 1995) that eclipsis, like lenition developed in two stages. The first stage, it is claimed, affected only the voiceless consonants *c-*, *t-*, *f-*;[36] this stage, it is argued, occurred before the introduction of Gaelic to Scotland. The second stage in the development of eclipsis (i.e. of the voiced consonants *b-*, *d-*, *g-*) was different in Irish and Scottish Gaelic. In particular, it is argued that the Irish eclipsis of the voiced stops, i.e. *b-*, *d-*, *g-* → *m(b)-*, *n(d)-*, *ng-* may never have existed in Scottish Gaelic. This argues for an early significant split between northern and southern Gaelic.[37] This hypothesis, although partly based on comparative reconstruction, is also supported by the available historical evidence. The main body of evidence for eclipsis in earlier forms of Scottish Gaelic comes from three different sources: (1) fossils contained in modern Scottish Gaelic itself; (2) the sixteenth-century Book of the Dean of Lismore; (3) the twelfth-century Book of Deer.

The orthography of the sixteenth-century Book of the Dean of Lismore is unconventional from a Gaelic point of view, being based on the orthography of Middle Scots (Meek 1989). Although unconventional in nature, the orthography of this manuscript provides a rare and invaluable insight to a variety of registers, both formal and informal, of Scottish Gaelic in sixteenth-century Perthshire. See Watson (1924) for details. The Book of the Dean on the whole does not show evidence of the Irish eclipsis of the stops *b-*, *d-*, *g-*.[38]

Jackson's conservative, and perhaps overly cautious, interpretation of the language of the twelfth-century Gaelic Notes contained in the Book of Deer provides 'four cases where the spelling is such as to suit the later, Sc. G., treatment of nasalisation' (Jackson 1972, 143). A less conservative reading of the texts provides at least a further possible seven instances. This also includes 'the curious *i bBidbin*' which according to Jackson 'cannot be explained on either Irish or Sc. G. lines'; he concludes that 'it is probably a mistake' (Jackson 1972, 143). In fact this 'curious spelling' and some others are clearly and naturally explained if we assume the existence of the Scottish Gaelic type of eclipsis during the twelfth century; *i bBidbin* with double *b* may well be an attempt to signify a voiced (eclipsed) *b* as opposed to a radical voiceless *b* (see Ó Maolalaigh, forthcoming).

In conclusion, the available evidence for eclipsis in earlier Scottish Gaelic argues for the eclipsis of the voiceless consonants *c-*, *t-*, *p-*, *f-* only. There is little convincing evidence for the eclipsis of the consonants *b-*, *d-*, *g-*. This state of affairs may reflect the first stage of eclipsis although it should be pointed out that it is not incompatible with the modern system of eclipsis in Scottish Gaelic. How, you may ask, does this correlate with the place-name evidence? If the Irish system of eclipsis once existed in an earlier period of Scottish Gaelic, then we might expect it to be reflected in the surviving corpus of place-names, because of the tendency, already noted, for place-names to retain fossilised earlier forms. In particular we would expect to find instances of the second stage of eclipsis, i.e. *b-*, *d-*, *g-* → *m(b)-*, *n(d)-*, *ng-*. Before we turn our attention to Scottish place-names, let us first consider a selection of Irish place-names for reasons of comparison.

Irish place-name evidence
Anglicised place-names in Ireland provide ample testimony for the Irish system of eclipsis with the notable exception of eclipsed *g-*. In fact, the evidence illustrates quite clearly that the eclipsis of *g-* is rarely if ever represented in the anglicised form of Irish place-names. There is no doubt a phonological (distributional) reason for this. English, though it has the velar nasal sound *ng* /ŋ/ does not use it word initially or at the beginning of a stressed syllable. The substitution of /g/ for /ŋ/ is therefore understandable in such cases.[39] Leaving aside sporadic instances of the non-eclipsis of some consonants, e.g. *b-* in Ballinabranagh (CLW), *p-* in Ardnapreaghaun (LMK), and *c-* in Ballynacleragh (DWN),[40] anglicised Irish place-names consistently reflect the eclipsis of *c-*, *t-*, *p-*, *f-*, and, significantly for us, also of *b-*, *d-*. The Irish evidence warns us against expecting instances of eclipsed *g-* in anglicised Scottish place-names but on the other hand encourages us to look for instances of eclipsed *b-* and *d-*.

Eclipsis shown:

Boola**na**ve	<	Buaile na **n**Dámh (KRY)
Drumna**na**liv	<	Druim na **n**Dealbh (MON)
Rena**n**irree	<	Rae na **n**Doirí (CRK)
Foil**na**man	<	Faill na **m**Ban (TPY)
Slieve**na**mon	<	Sliabh na **m**Ban (TPY)
Ardna**m**oghill	<	Ard na **m**Buachaill (DNL)
Dunna**ma**ggan	<	Dún na **m**Bogán (KLD, KLK)
Glena**g**eary	<	Gleann na **g**Caorach (DUB)
Dun**d**rod	<	Dún na **d**Trod (ANT)
Carrig**na**var	<	Carraig na **bh**Fear (CRK)

Eclipsis not shown:

Ballinabranagh	<	Baile na **m**Breatnach (CLW)
Ardnapreaghaun	<	Ard na **b**Préachán (LMK)
Ballinagore	<	Baile na **n**Gobhar (WMH)
Ballynagaul	<	Baile na **n**Gall (WFD)
Clonegall	<	Cluain na **n**Gall (CLW)
Cloneygowan	<	Cluain na **n**Gamhain (OFY)
Aughnagon	<	Achadh na **n**Gabhann (DWN)
Donegal	<	Dún na **n**Gall (DNL)

Scottish place-name evidence

Watson (1926, 240–3) discusses some of the place-name evidence for eclipsis in Scottish Gaelic, listing 35 apparent examples. He distinguishes between two types of place-names: (A) place-names which do not show any evidence of eclipsis in their current form but for which historical variant spellings exist which do apparently show evidence of eclipsis; (B)[41] place-names whose current forms apparently show evidence of eclipsis.

Watson's examples are as follows:

		Modern form	Gaelic	Earlier attestations
(A)				
	1	Kiltarlity	Cill Taraghlain	Gylltalargyn (1203/24)[42]
	2	Kilpeter	Cill Pheadair	Gillepedre (1362)
	3	*Kilcalmkill	Cill Chaluim Chille	Gillecallumkille (1566)
	4	*Kilchrist	Cill Chrìost	Gilzacrest (1496)
	5	Killiehangie	Cill Chaomhaidh	Gilliquhamby (1558) [note also *Kelchemi c.*1190]
	6	Bochastle	Both Chaisteil	**M**ouchester (1542)
	7	Bonskeid	Bonn Sgao(i)d	**M**onskeid (1511)
	8	Borenich	Both Reithnich	**M**ontrainyche (1508)
	9	Bunchrew	Bun Chraoibh(e)	**M**onchrwe (1507)
	10	Ardencaple	Ard na **g**Capull	Airdendgapill (1351)
	11	Mucomir	Magh **g**Comair	Mogomar (1500)

	Modern form	from Gaelic
(B1)		
	Moness	i **m**Bun Easa
	Munlochy	i **m**Bun Locha
	Meadarloch (Eng. Benderloch)	i **m**Beinn eadar D(h)à Loch
	Muckairn	i **m**Both Càrna
(B2)		
	Achnagairn	Achadh na **g**Carn

1: Place-names as a Resource for the Historical Linguist

Dalnagairn	Dail na **g**Carn
Achnagullan	Achadh na **g**Cuilean
Alt **G**ellagach	Allt na **g**Cealgach
Balnagore	Baile na **g**Corr
Balnaguard	Baile na **g**Ceard
Cairnagad	Carn na **g**Cat[43]
Fin**e**gand	Féith na **g**Ceann
Drumgoudrom	Druim **g**Colldroma
Dalnavert	Dail na **bh**Feart
Dail na **bh**Fàd	Dail na **bh**Fàd
Blairnavaid	Blàr na **bh**Fàd
Loinveg	Lòn na **bh**Fiodhag
Bada na **B**reasach	Bad na **b**Preasach
Dund**u**rcus	Dùn **d**Turcais (?)

Watson claims that instances of initial *g*- and *m*- in examples A(1–9) and B1 represent original eclipsed *c*- and *b*- respectively. He argues that such instances have arisen 'due to the influence of the preposition *in* or *an*' which we have already noted is an eclipsing particle. He argues that phrases like *i mBun Locha* meaning 'in Bun Locha' with eclipsed *b*- would have been 'in constant use' and would naturally have given rise to *Mun Locha (Munlochy) which eventually replaced the original name *Bun Locha*. Similarly *i gCill Pheadair* meaning 'in Cill Pheadair' with eclipsed *g*- would have given rise to *Gill Pheadair in place of *Cill Pheadair*. I am sceptical of this type of locative explanation in these cases.[44] We shall see below that there are other valid explanations which explain the alternations *c*- ~ *g*- and *b*- ~ *m*- in the above examples. Even if instances of *g*- for *c*- in A(1–5) do indeed represent instances of the eclipsis of *c*- to *g*-, they can be discounted for our purposes on the grounds that they are not inconsistent with the Scottish Gaelic type of eclipsis. It should also be noted that the association of *cill*- names with *giolla*- names in the minds of non-Gaelic speakers may also have been a contributory factor in the occurrence of *g*- for Gaelic *cill* in examples A(1–5).[45]

It has not hitherto been noted that the majority of Watson's examples for the apparent eclipsis of *b*- to *m*-, with the possible exceptions of *Meadarloch* and *Muckairn*, may in fact represent cases of generic element substitution and variation, involving the element *Mon*-, presumably from *Monadh*, or in some instances *Mòine* or perhaps even *Muine*.[46] It is perhaps of significance that the majority of these names contain *b* + vowel + *n*, which may have facilitated the substitution, as we shall see below. Further research into the occurrence of these place-names in the historical record would need to be carried out before any firm conclusions could be made, however. Taylor

(1995, 441, 475) observes that Borenich (A8 above) 'occurs with variation in the generic between *both* and *mòine* in the early sixteenth c[entury]'. He goes on to say that *Montrainyche* 'seems to show the equally common confusion between *mòine* and *monadh*'. See also Taylor (1996, 105).[47] Generic element substitution and variation, which is well documented in the case of *peit* and *baile* in Scotland, is discussed in detail by Taylor (1997).[48] He makes the important point that such pairs may not originally have referred to the same place but rather that they 'were used to refer to different parts or aspects of the same place'.

Furthermore, the alternation in the group of words which Watson claims to be evidence for the eclipsis of *b*- in Scottish Gaelic could just as well be explained as back-formations (made by listeners, not speakers!) based on the equivalence of the lenited initials *bh*- = *mh*- which would regularly occur in locative phrases involving the leniting preposition *do* 'to' and also in genitival phrases involving words such as *muinntir*, *daoine* 'people' and so on. The development in question may be illustrated with the place-name *Bun Locha*. When lenited in phrases such as *do Bhun Locha* 'to Bun Locha' or *muinntir Bhun Locha* 'the people of Bun Locha', it is possible that the equivalence *bh*- = *mh*- may have suggested an underlying *m*- rather than *b*- especially when we consider that that the majority of Watson's examples of the change *b*- > *m*- involve words of the shape *b* + vowel + *n*. The nasality of the vowel preceding the nasal consonant in such words (e.g. *Bun*) could have been a factor which reinforced the impression of an underlying lenited initial *m*-.[49] This also applies to the place-name Benderloch (*Beinn-eadar-dà Loch*) which occurs in Gaelic as *Meadarloch*. The disappearance of the *-n-* of Benderloch would seem to lend weight to the re-assignment of nasality from the nasal consonant to the initial consonant as argued above. At a popular etymological level, it is possible that *Meadarloch* may have been interpreted as containing the Gaelic word *meadar* which in modern Scottish Gaelic means 'wooden vessel' or 'measure'.[50]

Watson (1926, 241) derives the Lorne place-name Muckairn from *Both-càrna* '? hut of flesh', the question mark indicating a tentative translation; he explains the *m*- as a further instance of the eclipsis of *b*-. The evidence for this derivation comes from two poems written by the eighteenth-century poet Alasdair Mac Mhaighstir Alasdair, where the place-name is spelled *Bucàrna* and *Bocàrna* respectively (MacDonald 1924, 262, 324). The earliest records for this place-name, dating from the early sixteenth century, consistently show it with initial *m*-, spelled variously as *Mocarne* (1527, 1553), *Moukcarne* (1546), *Muckarn* (1561), *Muckairn* (1564), *Muckairne* (1606) (*Origines Parochiales Scotiae*, vol. 2, part 1, 132–4). These forms may suggest that the place-name derives from a Gaelic word with initial *m-* and that MacMhaighstir Alasdair (or somebody before him) may have been indulging in a bit of folk-etymologising when he wrote *Bu-càrna*.[51] One plausible derivation for Muckairn, although there is no historical evidence for it, would be **Magh Ca(i)rn* 'plain of the cairn(s)'. Cf. *Magh Comair* > Mucomir (Watson 1926, 241, 500).

Our preliminary discussion has called into serious question Watson's evidence for the Irish type of eclipsis of *b-* in the Scottish place-names listed under A and B1 above. In any case we have shown that most if not all of these examples are explicable in terms other than eclipsis. It is perhaps most likely that the change of *b-* to *m-* in Scottish place-names is due to a number of factors including the phonological shape of the place-name elements involved as well as generic element substitution and variation.

We turn our attention now to examples A(10–11) and B2 which incidentally represent the bulk of Watson's evidence, and are by far the most convincing examples for the existence of eclipsis in Scottish Gaelic. They all, with the exception of Mucomir (A11), Drumgoudrom (B2) and the dubious Dundurcus (B2), contain instances of the eclipsing genitive plural article *na(n)*. It will be clear at a glance that these examples provide evidence only for the eclipsis of the consonants *c-*, *t-*, *p-*, *f-*. The absence of any evidence for the eclipsis of the voiced stops *b-*, *d-* (*g-*) following the genitive plural article, when compared with the Irish place-name evidence, is noteworthy and surely significant.

We must conclude that Watson's place-name evidence for the Irish type of eclipsis in Scottish Gaelic provides convincing evidence for the eclipsis of the consonants *c-*, *t-*, *p-*, *f-* only, which as we have already noted, is consistent with the modern Scottish Gaelic type of eclipsis. A cursory survey of the evidence for eclipsis in Scottish place-names culled from place-name studies unavailable to Watson, concurs with the conclusion reached above, i.e. that clear evidence of eclipsis in Scottish place-names exists only for the modern Scottish Gaelic type of eclipsis. Galloway place-names provide a notable exception, however.[52]

GALLOWAY[53]

/k/ → /g/
Bengairn   < Beinn na(n) **g**Cairn
Damnagclaur < (i d)Tam na(n) **g**Clàr

/f/ → /v/
Dunveoch   < Dùn na(m) **bh**Fitheach
Benaveoch[54] < Beinn na **bh**Fitheach

Knocknavar	<	Cnoc na **bh**Fear

/b/ → /m/[55]

Drum**m**uddioch[56]	<	Druim (na) **m**Bodach
Dun**m**an	<	Dùn na **m**Ban
Barna**m**on	<	Barr na **m**Ban
Knock**m**an	<	Cnoc na **m**Ban
Lagni**m**awn	<	Lag na **m**Ban
Knockna**m**ad	<	Cnoc na **m**Bàd

/b/ → /b/

Auchna**b**ony	<	Achadh nam **B**anbh (?)

/g/ → /g/

Blairna**g**obber	<	Blàr nan **G**obhar
Ilana**g**uy	<	Eilean nan **G**éadh

Assuming that our derivations are correct, the Galloway evidence is quite unique in Scottish terms as it unequivocally points towards the Irish type of eclipsis of *b-* to *m-* and by implication implies that the Irish system of eclipsis may have been a feature of Galloway Gaelic. It is noteworthy that the Galloway evidence, unlike that of other areas in Scotland, provides many examples of the eclipsis of *b-* following the eclipsing genitive plural article. The Galloway evidence raises several interesting questions. In terms of the eclipsis of *b-*, Galloway appears to align with Irish and Manx. This suggests a geographical division of eclipsis on a north-south axis connecting Ireland, Man and Galloway but separating these areas from the rest of Scotland. Such an isogloss forces us to think in terms of Northern *versus* Southern Gaelic as opposed to Eastern *versus* Western Gaelic, a division which has been suggested independently elsewhere (Ó Buachalla 1988, 58). This isogloss, though explicable in terms of a linguistic continuum, nevertheless raises certain questions with regard to the date and moreover the origin of the settlement of Galloway. It also raises questions about the status of Manx in relation to Irish and Scottish Gaelic. Such considerations are, however, outwith the scope of the present chapter.[57]

3. *-in* endings

So far we have been considering place-names as a resource for the linguist. Now I would like to change tack slightly and illustrate how the linguist might act as a resource for place-name study. I propose to focus on a problem which has faced Scottish place-name scholars for some time now and for which a satisfactory explanation has yet to be advanced. The problem in question relates to modern place-names ending in *-ie* or *-y* whose earlier attested charter forms end in -*in*. These endings invariably occur in the syllable immediately following the nuclear stressed syllable. The situation may be summarised as follows.

The phenomenon has been observed mostly in eastern Scotland in the areas of earliest contact between Scottish Gaelic and Scots, e.g. in East Aberdeenshire and Fife, but also in West Lothian and in Stirlingshire, as the representative examples below illustrate.[58] Spellings with -*in*, -*yn* are common in twelfth and thirteenth century sources but beginning in the fourteenth (in some cases as early as the twelfth and thirteenth centuries) these endings are replaced with -*ie*, -*y* spellings. In most, if not all, cases it is the -*ie*, -*y* pronunciation and spelling which has survived to the present day. Alexander (1952, xxv), however, notes for Aberdeenshire that 'in a very few cases both -*ie* and the -*in* ending have been heard in actual use with the same name, the latter only with the old people; as Monaltrie and Monaltrin, Tamnaverie and Tamnaverin.' The question which faces us is how to explain the origin of these -*in* endings and also how to account for the change from -*in* to -*ie*.

ABERDEEN (EAST)[59]

Modern name	Earliest -in	Earliest -ie
Altrie	*Alteri(n)* (12th C, BDeer)	–
*Alde[60]	*Aldín/-in* (12th C, BDeer)	–
Biffie	*Bidbin* (12th C, BDeer)	*Biffy* (1544)
Bourtie	*Bourdyn* (c.1175)	*Bouharty* (1342)
Cairnbrogie	*Carrinbrogyn* (1234)	*Carnbrogy* (1611)

STIRLINGSHIRE[61]

Airthrie	*Athran, Atheran* (c.1200)	*Athray, Atheray* (1317)?
Kersie	*Carsyn* (1195)	*Karsy* (c.1150)

WEST LOTHIAN[62]

Craigie	*Cragin* (1178)	-*y* (1296–1693)
	-*yn* (1296)	-*ie* (1670)
Binny	*Bennyn* (c.1200)	*Benny* (1477)
	Benyn (c.1244)	*Binnie* (1586)
Dalmeny	*Dunmanyn* (1214–1471)	*Dumanie*[63] (c.1180)
		Dumany (1378)

FIFE[64]

	Latest -in	Earliest -ie
Kinglassie	*Kilglassin* (1235)	*Kynglassy* (c.1245)
Logie (by Dunfermline)	*Logine* (c.1260)	*Logy* (1506)
Pitlochie	*Petclochin* (c.1220)	*Petlochy* (1448)
Torry	*Torrin* (1231)	*Torry* (c.1250)
Kinaldy	*Kynnadin* (c.1220)	*Kynaldy* (1375)

Three explanations have been put forward to date:

a. -*in* spellings reflect a scribal convention for writing place-names which did not reflect phonological reality (Taylor 1995, 42).
b. -*in* is an oblique case ending, locative or genitive, originating in Old Irish *n*-stem nouns (Watson 1926, 263; MacDonald 1941, 5; Alexander 1952, xxv–vi; Taylor 1995, 42).[65]
c. -*in* reflects a diminutive suffix (Black 1946, liv–v; Alexander 1952, xxv).

Discussion
Even if instances of -*in* spellings did acquire the status of mere scribal convention, this explanation avoids the question as it does not explain the origin of such spellings.[66] In any case, Alexander's testimony to fluctuation between -*ie* and -*in* pronunciations in modern times seems to rule this out, unless of course it is argued that -*in* realisations stem originally from written sources. One problem with positing a locative ending in -*in* is that there is little or no evidence within Gaelic that the locative/dative *n*-stem ending was ever productive in Scotland. If -*in* spellings do in fact represent locative *n*-stem endings, then we must assume that this inflection was productive in Eastern Scotland in the twelfth and thirteenth centuries since all instances of its occurrence do not represent underlying original nasal stems. If so, we may have established an early isogloss separating eastern from western Scottish Gaelic, with eastern dialects having a productive nasal locative ending and western dialects having a productive dental locative ending.[67] In support of the spread of a productive nasal stem inflection, we may note that the productive plural allomorph {*an*} in Scottish Gaelic, according to one interpretation, can be traced back to Old Irish *n*-stems (Ó Buachalla 1988). However attractive such an isogloss may first appear, it must be said that our knowledge of Eastern Scottish Gaelic, such as it has survived into the twentieth century and recorded by the Gaelic Linguistic Survey in the 1950s, does not support the existence of a productive nasal stem inflection in these dialects in modern times at least.[68] Moreover, the fact that the development of Eastern Scottish Gaelic dialects has tended towards the collapse of the traditional case system, would seem to militate against the spread of a productive nasal inflection in these areas.

Another problem with positing the Gaelic locative -*in* ending which has not hitherto been alluded to is its occurrence with compound names consisting of noun + noun where the suffixation of a locative ending following the second noun would be unexpected in Gaelic, e.g. in *Carrinbrogyn, Kinglassin, Pitlochin, Kinaldyn*, etc. It is this fact which no doubt led Alexander (1952, xxv–vi) and Taylor (1995, 42) to posit a genitive *n*-stem ending. Furthermore those who have put forward the locative *n*-stem ending have failed to explain the change *-in* to *-ie*. One plausible explanation would of course be the adoption of the Gaelic locative dental stem ending *-idh* in place of *-in*. Alternatively, the loss of final *-n* in this group of place-names could well indicate the loss in Gaelic of a productive nasal stem inflexion in some areas as early as the thirteenth century. Another plausible explanation of the variation between *-in* and *-ie* forms could be that while *-in* forms may have reflected locative (or in some cases genitive) forms, *-ie* forms may have represented nominative uninflected forms with final vowels, final unstressed vowels in Gaelic frequently being represented by *-ie* in Scots.

It was George Fraser Black (1946, liv) in his magisterial book, *The Surnames of Scotland*, who first suggested that the *-ie* suffix in certain Scottish place-names represented 'the weakened or unstressed form of another common diminutive place-name ending *-in*'. He quotes the *Bidbin* example from the twelfth-century Book of Deer for later *Biffie*.[69] Black, however, offers no explanation for the diminutive suffix *-in* and it is unclear whether he took it to be a Gaelic or a Scots element. Alexander (1952, xxv) also noted the connection between earlier *-in* and later *-ie*. He notes: 'it would accordingly seem that *-ie* is, perhaps in most cases, a broken-down form of the ending *-in*', which he suggested may exhibit 'a formerly widespread [Gaelic] stem-ending'. Although he admits that many cases of modern *-ie* may reflect locatives, he notes: 'but by no means all of our *-ie* endings can be referred off-hand to that origin'. He goes on to say that 'other origins may have to be looked for in particular instances; as, for example, the ending *-an*, which has a diminutive, sometimes a plural, implication'. He concludes his treatment of the *-ie*, *-y* endings by saying that 'the subject awaits further study'.

I support the view that some at least of the *-in* endings which later appear as *-ie* are likely to derive from an original Gaelic diminutive suffix. I would like to suggest, as a contribution to the *-in* ~ *-ie* debate, that the diminutive in question may be the Gaelic suffix *-ín* and/or *-ḗin*. The use of diminutives in place-names is well-known[70] and the use of *-ín* is of course quite common in Irish place-names as the following examples illustrate: Balbriggan < *Baile Brigín* (DUB), Ballinkilleen < *Baile an Chillín* (CLW), Ballyandreen < *Baile Aindrín* (CRK), Derreen < *Doirín* (CLW), Dromin < *Druimín* (LMK, LTH), Gorteen < *Goirtín* (GLY, SGO, WFD, etc.), Kilcomin < *Cill Chuimín* (OFY), Killeen < *Cillín* (TPY).[71]

If we assume an underlying *-ín* /iːnʹ/, or possibly *-ḗin* /eːnʹ/ in some at least of our place-names for which -*in* spellings are attested in the twelfth and thirteenth centuries, I believe we can explain the development -*in* (-*éin*) > -*ie* in a straightforward fashion. The 'change' -*in* /iːnʹ/, -*éin* /eːnʹ/ to -*ie* /iː/, with loss of the final nasal is not remarkable in phonetic terms. Indeed it is quite plausible that the final nasal be lost following a phonetically long or high (front) vowel. Non-Gaelic speakers, when confronted with Gaelic -*in* /iːnʹ/, -*éin* /eːnʹ/ could conceivably have heard or detected a final (perhaps nasalised) vowel /iː/ or /ĩː/ with no final consonant. This impression may have been further reinforced by the presence already of the Scots diminutive ending -*ie* in these areas.[72] Alternatively the change -*in* /-*éin* > -*ie* could be explained as a direct translation of the Gaelic diminutive -*in* /-*éin* into the Scots diminutive -*ie*. This process would demand a high level of bilingualism which is undisputed during our period. The diminutive hypothesis put forward here fits well with -*in* endings being used with compound names of the shape noun + noun, where the diminutive ending would regularly be attached to the second noun. See *Carn Bróigín* below.

It is worth noting that there is little consensus among scholars with regard to the origin of the Scots diminutive -*ie*, -*y*, which is particularly common in north-eastern Scotland. Most agree that the suffix is first attested in the fifteenth century in Scotland (Sundén 1910, 136; Marchand 1969, 298). The *Oxford English Dictionary* (under -y suffix) explains it from names like Davy, Mathy which rendered Old French Davi, Mathé 'which have the appearance of being pet forms of David, Mathou'.[73] Marchand (1969, 298) correctly dismisses this explanation on the grounds that it is difficult to accept the spread of -*y* suffixes from words which were monomorphemic. Sundén (1910, 144, 163–4) argues that the suffix -*y* originally had no hypocoristic function but was extended from Old English (e.g. Wolsi), French (Mary) or Scandinavian (Aki) proper names with final -*i* and -*y* to monomorphemic elliptical hypocoristic names, e.g. Addy, Batty, Benny, Roby. He argues that it was the use of the -*y* suffix with hypocoristic forms which gave it its hypocoristic force. Jespersen (1933, 297) and Marchand (1969, 298) both repudiate this view. Both refer to the unlikely extension of a suffix from words which are not analysable as bimorphemic words. Jespersen suggests that -*y* could well be a reflex of the Middle English hypocoristic suffix -*e*.[74] However, he is more inclined to see the origin of hypercoristic [i] suffixes in the general symbolic value of high front vowels. Sapir (1929) proves experimentally the innate psychological reality of the correspondence between high front vowels, particularly [i], and small objects or smallness in general. Jespersen (1933, 283, first published in 1922), illustrates convincingly from a range of languages (mostly Germanic) that the high front vowel [i] 'serves very often to indicate what is small, slight, insignificant, or weak'. The symbolic hypothesis is accepted by Marchand (1969, 298–9) and Nieuwenhuis (1985, 192). The widespread use of [i] suffixes (or in some cases palatalisation, e.g. Basque) in diminutives throughout western and central Europe is noted by Nieuwenhuis (1985, 32, 126–7 *et passim*).

The use of diminutives in languages during certain periods is very frequently subject to the whims of fashion, see Nieuwenhuis (1985, 199, 207–10).[75] It is conceivable that an upsurge in the use of diminutives may come about as a result of contact between two languages. It is perhaps significant that the area in Scotland where the diminutive suffix *-ie*, *-y* is the most productive includes many of Scotland's earliest towns, burghs and ports[76] through which trading with the Low Countries, the Baltic and France from the twelfth to the sixteenth century was conducted.[77] Nieuwenhuis (1985, 210, n. 4) suggests a possible link between the Scots diminutive suffix *-y* and the Dutch diminutive suffix *-je*[78] arguing that the period of expansion in the use of the Scots suffix *-ie* coincides with the period of trade between Scottish and Dutch ports during the sixteenth and seventeenth centuries. Bratus (1969, 2) refers to the unlikely possibility of 'the custom of using diminutives' being brought to Aberdeen 'by one of the Slavonic tribes, the Wends, who would have reached the shores of Scotland and landed around Aberdeen (Bulloch)'. It is tempting to add Gaelic to the range of possibilities. It could well be that contact between Gaelic and Scots provided the impetus for an increase in use of the diminutive suffix *-ie*, *-y*.[79] We have already remarked on the phonetic similarity of Gaelic *-ín*, *-éin* and Scots *-ie*, *-y*. We draw attention below to the frequent occurrence of the diminutive suffix *-ín* (mostly hypocoristic) in twelfth-century Gaelic texts from the same area (i.e. the Gaelic notes contained in the Book of Deer).

Once the transition from Gaelic *-ín* to Scots *-ie* had taken place or at least once the alternation *-ín* ~ *-ie* had been established, it would have been but a small step to replace by analogy, whether in writing or in speech, *-ie* from other sources with *-ín* and vice versa. I offer this as yet another alternative explanation to the problem of *-ín* ~ *-ie* variation in Scottish place-names, which may partially explain the preponderance of *-ín* spellings in the early sources. The explanation put forward here would imply that the alternation between *-ie* and *-ín* (old speakers only) in some modern Aberdeenshire place-names suggests that *-ín* represents the original Gaelic pronunciation and *-ie* the 'newer' Scots pronunciation.

An underlying diminutive *-ín*, *-éin* seems to suit at least one, and perhaps more, of the names listed above for which earlier *-in* spellings are attested.

The West Lothian place-name Binny (MacDonald, 49) which appears c.1200 as *Bennyn*, may be very plausibly derived from Gaelic *b(e)innín* or *b(e)innéin* meaning something like 'a little peak or hill', which is in fact a perfect description of the place. As MacDonald (1941, 49) correctly points out, Watson's (1926, 146) derivation from the locative of *Binneach* is not supported by the earlier forms. Similarly, the West Lothian place-name Craigie (*Cragin* 1178) could well derive from *Craigín* or *Creigéin* 'little crag'. The East Aberdeenshire place-name Cairnbrogie in Tarves parish (*Carrinbrogyn* 1234) could derive from *Carn Bróigín.[80]

Although it is not immediately obvious from the modern orthography that the diminutive suffix *-ín, -éin* ever existed in Scottish Gaelic, it can nevertheless be shown to have existed at an earlier stage of the language; it is now represented, according to dialect, by *-ean* and *-ein* in modern Scottish Gaelic. If textual evidence for the existence of the diminutive suffix *-ín* in medieval Gaelic Scotland were needed, we need look no further than the twelfth-century Book of Deer, which contains a relatively high incidence of diminutive proper nouns. According to Professor Jackson's reading of the texts, there are at least seven instances of diminutive suffixes with personal names and one with the place-name Aldín Alenn (V, 8). The personal names are as follows: Cú Líi mac Batín (II, 5), Meic-Dubbacín (II, 13), Matadín (III, 8), Feradac mac Mal-Bricín (IV, 3), Mal-Girc mac Trálín (IV, 4), Bróccín (VI, 6),[81] Mal-[F]aechín (VI, 7).[82] It has not hitherto been noticed that at least four of these names occur in the manuscript with a stroke over the final vowel *-in*. These strokes were unfortunately excluded by Professor Jackson in his edition of the Gaelic Notes in the Book of Deer since he regarded their function as merely 'to indicate that the language is vernacular, not Latin – that is to say, they are used very much as we use italics' (Jackson 1972, 17). The possibility that some at least of these strokes may have indicated length cannot, however, be discounted.[83]

Our interpretation raises some important questions with respect to some of Jackson's readings in the Book of Deer, in particular, the place-name Biffie, which occurs as follows: *Bidbin* (accusative); *i bBidbin* (dative). Jackson (1972, 50) takes *Bidbin* to be an oblique *n*-stem ending of an underlying *Bidbe or more likely *Bithbe. The manuscript, however, has an acute accent over the final syllable in both occurrences, which may imply *Bidbín* with a diminutive *-ín* ending. If so, the change *Bidbín* to Biffie could be explained in the same manner as the change *Beinnín* to Binny suggested above.

The place-name Rathelpie, St Andrews FIF appears as *Rathelpin* in the late twelfth century (Taylor 1995, 419) and thus provides a further instance of the development *-in* > *-ie*. Taylor tentatively suggests that this place-name[84] and indeed another from Fife, namely Skelpie (< *Caskelpie*), contains the personal name Alpin. If this is correct and if the diminutive *-ín*, *-éin* hypothesis is accepted, it may imply that the underlying name in these place-names is *Alpín* or *Ailpéin*. In medieval Gaelic genealogical sources, the name frequently occurs as *Alpin* (presumably for *Ailpín*), see O'Brien (1976, 328, 426). However, the name occurs as *Ailpé(i)n* (in rhyme with *dé(i)n*) in the *Duan Albanach*, see Jackson (1956, 162, line 66). It should also be noted that the pronunciation of the surname *MacAilpein* in modern Scottish Gaelic supports an underlying *Ailpín* or *Ailpéin*, with an original long vowel in the second syllable.

It is unlikely, however, that all instances of *-in*, *-yn* ~ *ie*, *-y* variation in the early sources represent an underlying Gaelic diminutive, and another explanation for the development *-in*, *-yn* > *-ie*, *-y* is accordingly required. There are two possibilities. The change is a phonetic one, or alternatively it represents a case of morphological substitution or indeed morphological reduction as was argued earlier. Either explanation admits the possibility of the change occurring (a) internally within a particular language or (b) across language boundaries as a result of language contact phenomena. Let us first consider the possible phonetic explanations for the change. The loss of final *-n* is not attested in Gaelic, Scots, English or Anglo-Norman so far as I am aware.[85] That there was no universal change of *-in* > *-ie* is borne out by the numerous instances of place- and personal names where the final nasal is preserved, e.g. Elgin, Cummin(g), Jardin(e) etc. Sundén (1910, 160–1), however, notes alternation between *-in* and *-y* in four English names from the thirteenth and fourteenth centuries: Astin~Asty (1273), Derkyn~Derky (1379), Dewyn~Dewy (1379), Germyn~Jermy (*c.*1300).[86] Sundén stated that 'the *y*-forms are no doubt modifications of the *-in* forms' but it is not clear whether or not 'modifications' here refers to phonetic or morphological modifications. The only viable phonetic explanation of the pairs *-in*~*-y* seems to be an acoustic one which occurred across language boundaries. If the ending *-in* is taken to be Gaelic, then it is conceivable that final *-in* may have appeared acoustically to non-Gaelic speakers as [i(:)]. We have already suggested this in the case of the Gaelic diminutive suffix *-ín*. However, that this development may have applied to other Gaelic *-in* suffixes (e.g. locative/dative *-in*) is also possible. Alternatively, if we assume that *-in* is not a Gaelic suffix, it could be argued that *-in* spellings represent [i(:)] or [ĩ(:)], not [i(:)n]. This would imply the loss of final post vocalic nasals in Scots, English or Anglo-Norman, which we have already noted is improbable.

We now turn our attention to the morphological explanations for the change. We have already noted that the change -*in* > -*ie* could conceivably reflect a change in Gaelic from nasal to dental stem inflection (but not in those place-names of the structure noun + noun) or the loss of nasal stem declension altogether.[87] Alternatively -*in*, -*yn* could represent a non-Gaelic suffix which was substituted in a bilingual context for an underlying Gaelic [i(:)] suffix. An example of such a non-Gaelic suffix might be the French or Anglo-Norman diminutive -*in*,[88] the plausibility of which I leave for others to assess.

In Scotland the fact that the change -*in* > -*ie* appears to be only attested in place-names would seem to argue against the change being a phonetic one, unless of course we accept that place-names may be subject to different phonetic and phonological rules from ordinary lexical items (on which see below). The change is also attested in some Scottish surnames, most, if not all, of which, perhaps significantly, derive from place-name. Examples include: *de Clonin* (c.1214–18) ~ *de Cluny* (1263) Black (1946, 157) under Cluny; *de Dunlopyn* (c.1178–80) ~ *de Dounlopy* (1312) Black (1946, 230) under Dunlappie; *Gogyn* (1314) ~ *Gogy* (1330) Black (1946, 316) under Gogy; *Gortin* (c.1208) ~ *de Gorty* (1271) Black (1946, 320) s.v. Gorthie.[89] The restriction of the change to place-names argues quite strongly in favour of the change being a morphological rather than a phonetic one.

4. Final unstressed -*ach* in Scottish place-names

The borrowing of Gaelic place-names with final unstressed -*ach* into Scots raises a number of interesting questions for the linguist and the onomastician alike. The subject has been dealt with by Nicolaisen (1986), (1988), (1993), (1996) on various occasions. Nicolaisen (1986, 142–3), (1988, 26–7) classifies the four main attested developments of Gaelic -*ach* in Scots and English as follows: (1) -*o*, (2) -*o(c)k*, (3) -*och*, (4) -*ach*.[90] The first development -*ach* > -*o* occurs 'mostly in the east of Scotland' (1986, 142) and is evidenced in Balerno, Balmerino, Largo, Pitsligo, etc. The second development whereby the final velar fricative is replaced by the homorganic velar stop is attested mostly in southern Scotland. We find it in Dumfriesshire (Dalgarnock), Dunbartonshire (Balernock), Midlothian (Cammo recorded as *Cambok* in 1296), Fife (Cambo recorded as *Camboc* in 1171–4). The third development which implies rounding of the unstressed vowel before the velar fricative which is retained (in writing at least) occurs mostly 'in areas which remained Gaelic-speaking much longer and in which Scottish English, rather than Scots, frequently replaced Gaelic'. It occurs in Badenoch (INV), Garioch (ABD), Tulloch (ROS), Rannoch (PER) etc.[91] The fourth development whereby Gaelic -*ach* remains unchanged occurs frequently in Gaelic map names in areas where Gaelic is still spoken or was still spoken until relatively recently, e.g. The Cabrach (BNF), Coigach (ROS) etc.

As the developments *-ach* > *-och*, *-o(c)k*, *-ach* are fairly transparent, they will not be discussed further here. The rest of this chapter will be devoted to a discussion of the development *-ach* > *-o*. Nicolaisen (1988, 27) concludes 'that the development of *-ach* > *-och* > *-o* occurred in the earliest contact zones of Gaelic and Scots ... while the retention of the final fricative in *-och*, as well as of *-ach* itself, reflects later phases of contact or lack of contact altogether.' Nicolaisen (1986, 142) describes the development *-ach* > *-och* > *-o* in the historical sources as follows:

> On the whole, the earliest forms of the thirteenth and fourteenth centuries preserve the original Gaelic *-ach* faithfully, but from the fifteenth century on forms in *-och*, *-auch* and *-augh* are the rule, indicating that rounding of the vowel from *-a-* to *-o-*, possibly under the influence of the following velar fricative, had already taken place; in fact, this tendency had obviously already existed for a while, as some of the thirteenth-century forms for Balerno (MLO) show, but spellings without the final fricative do not seem to be much recorded before the fifteenth century and only become plentiful from the sixteenth century onwards.

He goes on to say:

> One might therefore say that when place names ending in *-ach* were first adopted by Scots in the Scottish east and north-east ... in the thirteenth century, they kept the unstressed suffix more or less intact but began changing it to *-och* almost immediately in isolated instances, and wholesale from the fifteenth century on; the loss of the final fricative, although recorded in the fourteenth century, becomes common in the sixteenth century. (1986, 142)

Here is a sample list of names which illustrate the development, taken from Nicolaisen (1986, 142–3; 1996, 279–87):

	-ch (earliest)	**-o (earliest)**
Balerno MLO	*Balhernoch* (1280)	*Ballerno* (1461)
Balmerino FIF	*Balmurinach* (c.1212) *Balymorynoch* (1459)	*Balmurino* (1423)
Pitsligo ABD	*Petslegach* (1408)	*Petslego* (1459)
Aberlemno ANG	*Aberlevinach* (1202) *Abbyrlemnoch* (1488)	*Abirlemno* (1466)
Stracathro ANG	*Strukathrach* (1178–90) *Stracatherauch* (1226–31)	*Strucathrow* (1435)
Balmanno KCD	*Balmannoch* (1459)	*Balmannov* (1448)

Balmanno PER	Balmanach (1420)	Balmanow (1421)
Fetteresso KCD	Fethiresach (1287)	Fetheressau (1204–11)
	Fechiressoch (1419)	
Haddo ABD	Haldouch (1189)	Haddow (1528–9)

Nicolaisen's analysis implies that the development *-ach* > *-o* is a phonological development within Scots. He implies that the velar fricative was retained when Gaelic place-names were taken over into Scots in the twelfth, thirteenth and fourteenth centuries but subsequently lost in the course of the sixteenth century with sporadic instances of the loss occurring in the fourteenth and fifteenth centuries. He implies an intermediary stage where the vowel preceding the velar fricative was rounded. This rounding, he claims, is witnessed in spellings such as *-auch*, *-augh*, *-ouch*, *-och* which occur frequently from the fifteenth century, with isolated instances occurring in the thirteenth century. Nicolaisen (1993, 310–11) develops his analysis further in his paper 'Scottish Place Names as Evidence for Language Change' where he postulates a further intermediate stage whereby the rounded vowel [ɔ] is lengthened to [o:] (*sic*) before the velar fricative [χ]. The same argument is presented in Nicolaisen 1996. He claims that this lengthening is necessary 'in order to reach the final destination [o:] because there is no evidence to suggest that the voiceless velar fricative [χ] was ever lost in Scotland after short [ɔ].'[92] He goes on to say that the change [ɔχ] > [o:χ] > [o:] 'was mainly triggered in the sixteenth century, probably as a result of a bilingual Scottish-Gaelic-Scottish English period or, perhaps more plausibly, of an early post-Gaelic one.'

There are a number of problems with Professor Nicolaisen's analysis. The proposed lengthening of [o], or any other vowel for that matter, particularly in the unstressed position, before the velar fricative is not attested, so far as I am aware, either in Scots or in Scottish English. The synchronic evidence from modern Scots dialects indicates that the final velar fricative has on the whole been retained in normal lexical items in Mid and Northern Scots, the dialect areas for which the change *-ach* > *-o* is alleged to have taken place. The picture is further complicated given that most if not all instances of retained final unstressed *-ach* in modern Scots are borrowings from Scottish Gaelic, as we shall see in the discussion below. Nicolaisen's analysis, if correct, would have important implications for the linguist as it implies that place-names as linguistic items are subject to different linguistic rules than normal lexical items. It is well known that proper names in Gaelic may differ grammatically from ordinary lexical items. For instance, unlike common nouns, genitive singular masculine personal names and non-appellative place-names are obligatorily lenited. See Oftedal (1956, 181–2), Hamp (1959, 58–9). Indeed Hamp (1959, 57) notes that 'for the most part, the category which we conventionally term *proper name*, or simply *name*, seems to be defined not at all (or to a very limited degree) by formal linguistic criteria'. It is another matter entirely, however, to suggest that a phoneme ('a contrastive sound') or a string of phonemes should develop differently in onomastic and ordinary lexical items. That the phonology of onomastic and lexical items may develop independently is stated explicitly by Nicolaisen on various occasions:

> the Gaelic suffix *-ach* ... became *-o* in place-names..., there is no need for us to conclude – indeed, we have no right to do so – that such a plausible, demonstrable phonological development also affected Gaelic words in *-ach* which may have found their way into the Scots vocabulary of the region. (Nicolaisen 1988, 23)

He notes elsewhere:

> Historically, because of their special nature, names very often follow their own traffic rules, and although they are, for the purposes of communication, embedded in language, both spoken and written, they are also often strangers to it, fossils, grains of sand that may or may not turn into pearls. (Nicolaisen 1988, 24)

The suggestion that a sound change should occur in the onomasticon but not in the lexicon, although perhaps not impossible in linguistic terms, jars slightly with the linguist on scientific grounds. As it happens, the change *-ach* > *-o* is not completely restricted to the onomasticon. Instances from the lexicon, though not numerous, include *blatho* < *blàthach*, *cowdow/coddow* < *cullach*, *cupno(w)* < **colp(a)nach*, *kyloe* < *Gàidhealach*, *clairschow* < *clà(i)rs(e)ach*, *gluntow* < *glù(i)nt(e)ach* (on which see further below).[93] The proper names *Murdo* < *Muireadhach* and *CollKitto* < *Coll Ciotach*, provide further instances of the change *-ach* > *-o*, although strictly speaking these words belong to the onomasticon rather than to the lexicon. These considerations demand that a different solution be sought which will explain the facts more satisfactorily.

If we begin with the premise that the change *-ach* > *-o* is not a phonological development within Scots, then we can explain Scottish place-names in *-o* from Gaelic *-ach* in a different way. A cursory survey of Scots word phonology shows that final unstressed velar fricatives did not (normally) occur in the native lexicon of Scots.[94] In other words the phonological structure of Scots did not tolerate velar fricatives in the unstressed position. In such a scenario, Gaelic final *-ach* could conceivably have been borrowed without the final velar fricative. The adoption of Gaelic -*ach* into Scots as [o] with final rounded vowel to represent the velar element would be quite natural in both phonetic and phonological terms. Indeed the borrowing of Gaelic -*ach* as -*ock*, which occurs mostly in southern Scotland, where the velar element is retained as a stop rather than as a fricative seems to lend support to our hypothesis that final unstressed velar fricatives were not a feature of Scots phonology. In such cases Gaelic final unstressed *ch* has been replaced by the nearest Scots phonological equivalent.[95]

How does this hypothesis correlate with the recorded historical evidence? A round vowel is attested in some instances from the thirteenth century, with and without the final -*ch* (e.g. *Camehou* 1294 for modern Cambo, *Fetheressau* 1204–11 for Fetteresso, *Kirkintillo* 1287 for Kirkintilloch, *Balhernoch* 1280 for Balerno, see Nicolaisen (1986, 142; 1996, 279–87) for further examples), although Nicolaisen (1993, 310) states that 'there is some indication that it may have occurred as early as the twelfth [century] and been latent for several hundred years'. I would add that the existence of forms with final vocalic [o] in Scots' pronunciations of Gaelic words is not incompatible with -*ach*, -*auch*, -*augh*, -*och* spellings in the historical record. In other words, spellings like -*auch*, -*augh*, -*och*, rather than representing [oːχ] as suggested by Nicolaisen (1993, 310), could well have stood for [o] with 'silent' -*ch*, -*gh*.

Let us now consider Gaelic words with final -*ach* in the normal lexicon which have been borrowed into Scots. From the sixteenth century to the present century, final -*ach* in such words is consistently retained in Scots sources as the following examples illustrate, the numbers in brackets referring to the century in which the word is first attested: *bourach* (19) < *buarach*; *claddach* (19) < *cladach*; *closhach/clossach* (19) < *closach*; *gralloch* (19) < *greallach*; *greeshoch* (19) < *grìosach*; *ablach* (18) < *ablach*; *bladdoch* (18) < *blàthach*; *coronach* (18) < *corranach*; *pibroch* (18) < *pìobaireachd*; *Sassenach* (18) < *Sasannach*; *quigrich* (18) < *coigreach*; *bledoch* (16) < *blàthach*; *currach/currok* (16) < *curach*; *dorlach* (16) < *dorlach*; *enache* (16/15) < *eineach*; *larach* (16) < *làrach*; *toiseach* (16) < *toiseach*; *davach/dauch* (16) < *dabhach*. However, the following words illustrate the development -*ach* < -*o*:

bladdoch < blàthach

bladdoch (18/19), *bledoch* (16–20), **blatho** (19/20).

clarsach < clà(i)rs(e)ach

clarsach (20), *clarschach* (15–19), **clareschaw** (15/16), **clairschow** (16/17), *clerscha* (15–17).

colpindach* < **colp(a)nach

colpindach (11–17), *copnoche* (16), *coupnocht* (16), *cupno(w)* (16), *coupnay* (16/17).

***cuddoch* < *cullach*[96]**

cuddoch (18–20), **cowda** (18–20), *coldoch* (16), *cowdach* (16/17), *coddoch* (16/17), *kowdoch* (16/17), **cowdow** (16–17), **coddow** (16).

gluntoch* < *glù(i)nt(e)ach (?) (see footnote 93).

gluntoch (16), *gluntow* (15).

kyloe* < *Gàidhealach

kyloe (18–20), *kylie* (18–20).

It is clear from the above examples that Gaelic *-ach* is represented by *-o(w)* in a small number of words in Scots sources from the fifteenth to the present century. It may be significant that many of our examples of the change *-ach* > *-o(w)* belong to the fifteenth and sixteenth centuries. Rather than providing a date for the change *-ach* > *-o*, this evidence may support our suggestion that Gaelic *-ach* was borrowed without the final fricative in the earliest stratum of Gaelic borrowings. It is conceivable, indeed highly probable, that place-names would be among the earliest borrowings where contact occurred between Gaelic and Scots. In this respect, we may recall that in place-names, the development of *-ach* > *-o* 'occurred in the earliest contact zones of Gaelic and Scots' (Nicolaisen 1988, 27). Many of the Gaelic words in Scots which retain the final velar fricative, if borrowed early, may originally have been borrowed without the velar fricative. However, continued and renewed contact with Gaelic speakers may have helped to reinforce or reintroduce the final velar fricative in such words. Variation between *-ach/-och* and *-o(w)* in Scots words from Gaelic *-ach*, if they do not represent dialectal variation, may well support this possibility. If the derivation of *kyloe* from *Gàidhealach* is correct, the argument presented here would suggest that this borrowing occurred long before its first attestation in eighteenth-century Scots sources. If, on the other hand, *kyloe* is a late borrowing from *Gàidhealach*, the *-o* form may imply literary or learned forces at work.

Our discussion of Gaelic *-ach* in Scottish place-names has highlighted the difficulties which face the linguist and the onomastician alike when dealing with the historical onomasticon. The hypothesis presented here to account

1: Place-names as a Resource for the Historical Linguist

for *-och, -o* spellings draws particular attention to the highly conservative nature of that onomasticon, which from the linguist's point of view it would be treacherous to ignore.

The present chapter illustrates how place-names may be exploited as a resource by the historical linguist and equally how the linguist may provide an input to place-name studies. It highlights the rewarding and mutually beneficial nature of an interdisciplinary approach to the study of place-names. We have seen how place-names can provide a valuable window on the past where other sources are not so illuminating. This chapter merely scratches the surface, however. Much work remains to be done. When the results of the place-name projects in Edinburgh and St Andrews come to fruition, new avenues of research will open up and will enrich a wide variety of disciplines, particularly historical linguistics.

Abbreviations
For county abbreviations, see Abbreviations p. xiv.

Bibliography
Alexander, W.M., *The Place-Names of Aberdeenshire* (Aberdeen, 1952).
Barrow, G.W.S., 'Some East Fife Documents of the Twelfth and Thirteenth Centuries', in *The Scottish Tradition*, ed. G.W.S. Barrow (essays in honour of Ronald Gordon Cant) (1974), 23–43.
Black, G.F., *The Surnames of Scotland: Their Origin and Meaning* (New York, 1946; reprinted by Birlinn with new 'Amendments and Additions', 1993).
Black, R., Review of Jackson (1972), *Celtica,* 10 (1973), 264–7.
BorgstrPm, C. Hj., *The Dialects of the Outer Hebrides* (Oslo, 1940).
Bratus, B.V., *The Formation and Expressive Use of Diminutives*. Studies in the Modern Russian Language (London, 1969).
Breatnach, L., 'An Mheán-Ghaeilge', in McCone (1994), 221–333.
Crawford, B.E., *Scotland in Dark Age Britain* (Aberdeen, 1996).
Dieth, E., *A Grammar of the Buchan Dialect (Aberdeenshire): Descriptive and Historical* (Cambridge, 1932), vol. 1: Phonology and Accidence.
Dorian, N.C., *East Sutherland Gaelic* (Dublin, 1978).
Dorian, N.C., *Language Death* (Philadelphia, 1981).
Elcock, W.D., *The Romance Languages* (London, 1960; new and revised edn 1975).
Flanagan, D. and Flanagan, L., *Irish Place Names* (Dublin, 1994).
Greene, D., Review of Jackson (1972), *Studia Hibernica,* 12 (1973), 167–70.
Greene, D., 'Distinctive plural forms in Old and Middle Irish', *Ériu,* 25 (1974), 190–9.
Hamp, E., 'Proper Names in Scottish Gaelic', *Names,* 7.1 (March 1959), 57–9.
Hogan, E., *Onomasticon Goedelicum* (Dublin, 1910; reprinted by Four Courts Press, 1993).
Holmer, N.M., *The Gaelic of Arran* (Dublin, 1957).
Hughes, A. and Hannan, R.J., *Place-Names of Northern Ireland,* vol. 2. *County Down II:*

The Ards. The Northern Ireland Place-Name Project, general ed. G. Stockman (Belfast, 1992).

Irving, J., *Place Names of Dunbartonshire* (Edinburgh, 1928).

Jackson, K.H., 'The poem A Eolcha Alban uile', *Celtica,* 3 (1956), 149–67.

Jackson, K.H., *The Gaelic Notes in the Book of Deer* (Cambridge, 1972).

Jackson, K.H., *Common Gaelic: the Evolution of the Gaelic Languages,* Sir John Rhys Memorial Lecture (London, 1951). [Also in *Proceedings of the British Academy,* 37 (1951).]

Jespersen, O., 'Symbolic value of the vowel I', *Linguistica*: Selected Papers in English, French, and German (Copenhagen, 1933), 283–303.

Johnston, J.B., *The Place Names of Stirlingshire,* 2nd edn (Stirling, 1904).

MacAulay, D., Review of Jackson (1972), *Scottish Historical Review,* 54 (1975), 84–7.

MacDonald, A. and MacDonald, A., *The Poems of Alexander MacDonald* (Inverness, 1924).

MacDonald, A., *The Place-Names of West Lothian* (Edinburgh, 1941).

Marchand, H., *The Categories and Types of Present-day English Word-Formation,* 2nd edn (1969).

Maxwell, H., *The Place Names of Galloway: Their Origin and Meaning Considered* (Glasgow, 1930).

McCone, K. *et al.* (eds.), *Stair na Gaeilge* (Maynooth, 1994).

McManus, D., 'An Nua-Ghaeilge Chlasaiceach', in McCone (1994), 335–445.

Meek, D., 'The Scots-Gaelic scribes of late Medieval Perthshire: an overview of the orthography and contents of the Book of the Dean of Lismore', in *Bryght Lanternis: Essays on the Language and Literature of Medieval and Renaissance Scotland,* eds. J.D. McClure and M.R.G. Spiller (Aberdeen, 1989), 387–404.

Menger, L.E., *The Anglo-Norman Dialect: a Manual of its Phonology and Morphology* (London, 1904).

Nicolaisen, W.F.H., 'Gaelic Place Names in Scots', *Scottish Language,* 5 (1986), 140–6.

Nicolaisen, W.F.H., 'Gaelic and Scots 1300–1600: Some Place-name Evidence', in *Gaelic and Scots in Harmony,* ed. D. Thomson (Glasgow, 1988), 20–35.

Nicolaisen, W.F.H., 'Scottish Place Names as Evidence for Language Change', *Names,* 41 (1993), 306–13.

Nicolaisen, W.F.H., 'Gaelic *-ach* > Scots *o* in Scottish Place Names', *Scottish Gaelic Studies,* 17 (1996), 278–91.

Nieuwenhuis, P., 'Diminutives' (unpublished Ph.D. thesis, University of Edinburgh, 1985).

Ó Buachalla, B., 'MacNeill's Law and the plural marker -(e)an', *Proceedings of the Royal Irish Academy,* 88 C 3 (1988), 39–60.

Ó Cuív, B., 'The linguistic training of the Mediaeval Irish poet', *Celtica,* 10 (1973), 114–40.

Ó Cuív, B., 'A Mediaeval exercise in language planning', in *Progress in Linguistic Historiography,* ed. K. Koerner (Amsterdam, 1980), 23–34.

Oftedal, M., *The Gaelic of Leurbost* (Oslo, 1956).

Ó Maolalaigh, R., 'The Development of Eclipsis in Gaelic', *Scottish Language,* 14/15 (1995/96), 158–73.

Ó Maolalaigh, R., 'The language and orthography of the Gaelic Notes (in the Book of Deer), (provisional), in *Studies in the Book of Deer* (provisional title), ed. K. Forsyth (forthcoming).

Ó Maolfabhail, A., 'An Logainm *Tóin re Gaoith*', *Seanchas Ard Mhacha* (Armagh, 1982), 366–79.
Ó Sé, D., 'The forms of the personal pronouns in Gaelic dialects', *Éigse*, 29 (1996), 19–50.
O'Brien, M.A., *Corpus Genealogiarum Hiberniae* (Dublin, 1976; first published 1962).
O'Rahilly, T.F., 'Notes on Middle-Irish pronunciation', *Hermathena*, 20 (1930), 152–95.
O'Rahilly, T.F., *Irish Dialects Past and Present* (Dublin, 1932; reprinted with indexes 1972, 1976).
Pokorny, J., 'Da- in Irischen Ortsnamen', *Zeitschrift fur Celtische Philologie*, 14 (1923), 270–1.
Sapir, E., 'A study in phonetic symbolism', *Journal of Experimental Psychology*, 12 (1929), 225–39.
Sundén, K.F., 'On the origin of the hypocoristic suffix -y (-ie, -ey) in English', *Sertum Philologicum Carolo Ferdinando Johansson oblatum* (Göteborg, 1910), 131–70.
Taylor, S., 'Settlement Names in Fife' (unpublished Ph.D. thesis, University of Edinburgh, 1995).
Taylor, S., 'Place-Names and the Early Church in Eastern Scotland', in *Scotland in Dark Age Britain*, ed. B.E. Crawford (Aberdeen, 1996), 93–110.
Taylor, S., 'Generic element variation with special reference to Eastern Scotland', *Nomina*, 20 (forthcoming, 1997).
Thomson, R.L., 'The emergence of Scottish Gaelic', *Bards and Makars* (1977), 127–35.
Thurneysen, R., *A Grammar of Old Irish* (Dublin, 1946; reprinted 1993).
Toner, G. and Ó Mainnín, M.B., *Place-Names of Northern Ireland*, vol. 1. *County Down I: Newry and South-west Down*. The Northern Ireland Place-Name Project, general ed. G. Stockman (Belfast, 1992).
Watson, W.J., 'Vernacular Gaelic in the Book of the Dean of Lismore', *Transactions of the Gaelic Society of Inverness*, 31 (1924), 259–89.
Watson, W.J., *The History of the Celtic Place-Names of Scotland* (1926; reprinted by Birlinn, 1993).

Notes

1. I would like to thank Professor William Gillies and Drs Simon Taylor, Brian Ó Curnáin, Thomas Clancy and John MacInnes for supplying valuable comments and input to an earlier version of this chapter. The usual disclaimer applies.
2. Recent research into the historical development of the Gaelic languages has shown that Jackson's date for the break-up of Common Gaelic is too late. See Jackson (1951), Ó Buachalla (1977, 1988), Ó Maolalaigh (1995/96).
3. If the testimony of the seventeenth-century Annals of the Four Masters is accepted, it would appear that Scottish Gaelic was perceived by the Irish as being different from Irish Gaelic from at least the middle of the thirteenth century. The Annals of the Four Masters record that when Domhnall Ó Domhnaill of Donegal returned from Scotland in 1258, he addressed the messengers of Ó Néill through the medium of Scottish Gaelic, 'tria san nGaoidhilcc nAlbanaigh'. (Jackson 1951, 92), (O'Rahilly 1932, 162).
4. O'Rahilly (1932, 16) notes that evidence for the growth of dialects in Irish 'previous to the middle of the fifteenth century ... is very scanty indeed.' Leaving aside the twelfth century Book of Deer which is discussed below, the sixteenth century Book

of the Dean of Lismore is the first major source which bears direct testimony to the emergence of local dialects in Scottish Gaelic. See also Colm Ó Baoill's discussion of the evidence for Scotticisms in a fifteenth century Gaelic manuscript in 'Scotticisms in a manuscript of 1467', *Scottish Gaelic Studies*, XV (Spring, 1988), 122–39.

5. The place-name Rosneath which Watson (1926, 246) derives from Gaelic *Rosneimhidh* was presumably 'borrowed' when Gaelic *dh* was still pronounced as a dental.
6. See particularly the discussion below of Gaelic *-ach* in Scottish place-names.
7. Also *Torr na Gaoith*. See A. Cameron, *Arran Place Names* (Inverness, 1890), 12.
8. Ó Maolfabhail (1982, 366) claims to have found no instances in the Isle of Man. I am grateful to Art Ó Maolfabhail for supplying me with a copy of his article 'An Logainm *Tóin re Gaoith*'.
9. *Gaoth* meaning 'water' is attested in some Irish place-names, e.g. *Gaoth Dobhair* (Gweedore, DON), *Gaoth Sáile* (Gweesalia, MYO), *Gaoth Beara* (Gweebarra, DON).
10. But cf. *frae, fru liom* etc. in T. de Bhaldraithe, *Gaeilge Chois Fhairrge* (Dublin, 1953, 1977), 143.
11. For a full list of such names by county with forms from the historical record, see Ó Maolfabhail (1982, 376–8).
12. *Tóin re Gó* occurs in Sligo, with the element *gó* meaning 'sea (water)'. See *Dictionary of the Irish Language* (Royal Irish Academy, Dublin, 1983, 1990) under *gó*, p. 367. See also Ó Maolfabhail (1982, 369–70).
13. There is some fluctuation between *-l-* and *-r-* forms in the historical record of current *-r-* forms also. See Ó Maolfabhail (1982, 376–8).
14. It is possible that individual instances of *-l-* for *-r-* may represent cases of analogy on behalf of scribes and the recorders of place-names. Cf. Tonlegee (RSC), which appears in the published Ordnance Survey map with Tonerege, which appears in the unpublished version prepared in the years 1826–42. Ó Maolfabhail (1982, 376–7).
15. Sixteenth- and seventeenth-century Irish literary sources show much vacillation between the two prepositions *re* and *le*. See N. Williams 'Na canúintí a theacht chun solais', in *Stair na Gaeilge* (Maynooth, 1994, 462) ed. K. McCone *et al.*
16. Sporadic instances do occur in the modern languages, however. The non-lenition of *mór* in the Scottish Gaelic phrase *tìr-mór* 'mainland', if it does not derive from an accusative form, may reflect the original neuter gender of *tìr*. We may compare the Irish place-name Leitir Móir (GLY) lit. 'Big Hillside' which also contains the word *tír* (Leitir < *leath + tír*). Original neuter nouns frequently show fluctuation of gender in the modern dialects, e.g. *ainm* (m/f) 'name', *muir* (m/f) 'sea', etc.
17. See Watson (1926, 242–3). However, Invernauld and Cumbernauld are problematical as the *-n-* in both instances could derive from the *-n-* of the genitive singular or plural article *an, nan*.
18. But now Loch nan Eala with the genitive article.
19. But Maxwell (1930, 201) derives this name from *Loch an Eilergin*.
20. This is Watson's (1926, 242) derivation. Alexander (1952, 263) prefers *Drum Gauldrum* (sic) 'ridge of the shoulder ridge'. Note that the similar Kingoldrum ANG appears in the twelfth and thirteenth centuries as *Kingoueldrum*, etc. (e.g. *RRS* ii no.197), and seems to contain Gaelic *gobhal* 'fork' (*pace* Watson 1926, 242).

21. Watson (1926, 242).
22. Some northern dialects have articles of phonological shape /(ə)Nd/, but such forms are, in my view, more likely to be innovations rather than direct reflexes of Old Irish *ind*; the final /d/ of such forms signals word boundaries and is found in forms of the article where it would not be expected historically. See C. Hj. Borgstrøm, *The Dialects of the Outer Hebrides* (Oslo, 1940), 22–3, 78–80; M. Oftedal, *The Gaelic of Leurbost, Isle of Lewis* (Oslo, 1956), 207.
23. For the derivation of *-ruel* from *ruadhail*, see Watson, 1926, 473–4.
24. In the following examples, instances of *-nd-* could conceivably represent a Scots feature. They could represent examples of epenthetic *-d-* in Scots, though this seems unlikely since this development is uncommon in Scots. See R. Zai, *The Phonology of the Morebattle Dialect*, (Lucerne, 1942), 195–7; Dieth (1932, 123). Cf. C. Jones in *A History of English Phonology* (London, 1989), 122, who notes that instances of epenthetic *-d-* following *-n-* are rare, even in Middle English. Alternatively, *-nd-* could be a Scots graphemic representation of phonetic [n]; historical *-nd-* has usually been reduced to *-n-* in most Scots dialects. See Dieth and Zai ibid. Professor W. Gillies suggests to me that the *d* in these forms might represent a Scots interpretation of the Gaelic dental nasals which occur in forms of the article.
25. Cf. Jackson (1951, 89, n. 2).
26. Jackson's derivation implies the loss of intervocalic *-bh-* by the thirteenth century and perhaps also the reduction of *abhainn* to a monosyllable. However, it could be argued that a disyllabic *abhainn* (with silent *-bh-*) may have been perceived by non-Gaelic speakers as a monosyllable.
27. Note that Gaelic *fh-* is (usually) silent.
28. Watson (1926, 137–8) derives Ballindean from Gaelic *Baile an Deadhain* 'Farm of the Dean'. Although this interpretation may represent a genuine late interpretation of the name, it is not supported by the earlier forms. Nicolaisen (1976, 141) accepts Watson's derivation.
29. It is interesting in this context to note that the adverb which is used in many Scottish Gaelic dialects for 'downwards' *a-bhàn* contains the word *fàn* originally meaning 'slope'. See N.C. Dorian, *East Sutherland Gaelic* (Dublin, 1978), 203; M. Ó Murchú, *East Perthshire Gaelic* (Dublin, 1989), 292; H.C. Dieckhoff, *A Pronouncing Dictionary of Scottish Gaelic* (Glasgow, 1992), 1.
30. For similar instances of lenited proper names, although restricted to certain syntactic environments, in late twelfth century Irish literary sources, see Breatnach (1994, 237).
31. One gets the impression that the lenition is more frequently represented in scotticised and anglicised place-names in Scotland than in Ireland. This may well be due to the manner in which Irish place-names were recorded by the Ordnance Survey. For recent accounts of the Ordnance Survey and the scholars behind the scenes, see A. Ó Maolfabhail, 'An tSuirbhéireacht Ordanáis agus Logainmneacha na hÉireann 1824–34', *Proceedings of the Royal Irish Academy*, 89 (1989), 37–66; 'Éadbhard Ó Raghallaigh, Seán Ó Donnabháin agus an tSuirbhéireacht Ordanáis 1830–4', *Proceedings of the Royal Irish Academy*, 91 (1991), 73–104.
32. Auchencairn is problematical since it may contain the genitive plural article *nan* (or be a generalised Auchen- name coined outside a Gaelic milieu?). Genitive singular feminine proper names are not lenited in most Scottish Gaelic dialects (but see Hamp

1959, 59) although we might expect lenition historically in some cases. This may explain the absence of lenition in Kilbride from Gaelic *Cill Brighde* (several).
33. Some of the Shankill names in Ireland may derive from *sean* + *coll* 'hazel' or *coill* 'wood'. But see Flanagan (1994, 51).
34. The eclipsed sounds are usually written *gc-, dt-, bp-, bhf-, ng-, nd-, mb-* respectively.
35. There are three main types of eclipsis in Scottish Gaelic dialects. See Ó Maolalaigh (1995, 159–60) for details and references.
36. /p/ did not exist at this stage except in Latin and British loanwords and perhaps also in certain back formations e.g. *piuthar.*
37. The orthodoxy, however, speaks in terms of eastern vs. western Gaelic. See Jackson (1951).
38. I have noted a very small number of examples which appear to reflect the Irish system of eclipsis but these are rare exceptions and could be attributed to high register or Irish pronunciations.
39. However, the manner in which Irish Gaelic names were transferred to Ordnance Survey Maps in the nineteenth century cannot be discounted here. See note 31 above.
40. *Ballynacleragh* in 1435 for Irish *Baile na gCléireach* (Toner 1992, 72).
41. I have not listed Monzie, Monzievaird, Invernauld, Cumbernauld here as they are irrelevant to the present discussion.
42. All references from Watson (1926).
43. Watson (1926, 242) also gives *Lochan na gCat* from Perthshire.
44. T.F. O'Rahilly in his book *Early Irish History and Mythology* (Dublin, 1946), 373–4 invoked this type of locative explanation to derive the place-name *Ner* (which occurs in the Annals of Ulster s.a. 679, 623) from *Déar* 'Deer' in Aberdeenshire. This view has been accepted by a number of scholars including B.T. Hudson, 'Kings and Church in Early Scotland', *Scottish Historical Review* 73 (1994), 146–7; Jackson (1972, 6). *Ner* has, however, more recently and more plausibly been identified with the second element in the medieval parish Fetternear (*Fethirneir* 1157, 1163). See T.O. Clancy, 'Scottish Saints and National Identities in the Early Middle Ages', *Local Saints and Local Churches,* eds. R. Sharpe and A. Thacker (O.U.P., 1998, forthcoming). I am grateful to Dr Clancy for providing me with a copy of his forthcoming article.
45. Compare Gilchriston (ELO), Gilmerton (MLO) all of which derive from *giolla* (Watson, 1926, 134).
46. The *u* of *Mouchester* (1524), *Moucastell* (1579) (Watson 1926, 240) could be scribal errors for *Monchester* etc. with the common confusion of paired minims.
47. Watson (1926, 116) also derives the personal name Munro from Gaelic *Bun-rotha* 'by eclipsis of *b* after the preposition *in*'. Macbain, who had earlier made a similar suggestion, deriving it from *Bun-Ruadh*, nevertheless seems to have preferred *Monadh-Ruadh*. See 'Amendments and Additions', page 833 in the 1993 reprint of Black (1946).
48. Cf. Taylor (1995, 61–6).
49. Professor E. Hamp discusses the hearer's reassignment of nasality in Gaelic in syllables containing labial fricatives (*bh, mh*) and nasal segments in a number of articles. See E. Hamp 'St Ninian/Ronyan again', *Celtica*, 3 (1956), 290–4 (esp. 294);

1: Place-names as a Resource for the Historical Linguist

'Varia I', *Ériu*, 21 (1969), 87–8 (esp. 88); 'Scottish Gaelic *morair*', *Scottish Gaelic Studies*, 14 (1986), 138–41 (esp. 138); 'Varia' (esp. 5. *máiríon*) *Scottish Gaelic Studies*, 16 (1990), 191–5.

50. We may compare the use of *dabhach* 'vat etc.' in place-names (Watson 1926, 184–5, 235–6).
51. I owe this suggestion to Dr T. Clancy.
52. Compare the following place-names: East Aberdeen: Auchnagorth < *Achadh na(n) gcoirthe*; Auchnabo (*Auchnabo* (1612)) < *Achadh nam bó*; Lenabo (*Lennaboe* (1641)) < *Lèana nam bó* (Alexander 1952); Dunbartonshire: Ardencaple (*Ardengappil* (1351)) < *Aird na(n) gcapall* (?); Auchgower < *Achadh nan gobhar* (?) (Irving 1928). Indepth studies will, of course, need to be carried out throughout Scotland to test the validity of this preliminary conclusion, which is outwith the limits and remit of the present chapter.
53. Examples including Gaelic forms from Maxwell (1930).
54. Dunveoch and Benaveoch may contain the element *beathach, beithidheach, beothaich* 'beast, animal'.
55. Maxwell (1930) may not have been correct in suggesting an underlying *ban* (genitive plural of *bean* 'woman') in all instances for Dunman, Barnamon, Knockman and Lagnimawn. The long vowel which is implied by the English spelling of the final syllable in Lagnimawn does not fit comfortably with an original *ban*. MacQueen in 'The Gaelic Speakers of Galloway and Carrick', *Scottish Studies*, 17 (1973), 28, derives Dunman from Gaelic *Dùn na mBeann* 'fort of the peaks' or 'fort of the gables' which, if correct, would also support the Irish type of eclipsis of *b*-. Some of these examples could conceivably contain the element *Man* in 'Isle of Man'. However, the internal -*n*- in Barnamon, Lagnimawn, Knocknamad imply the presence of the article. Cf. Slievenamon < *Sliabh na mBan* (TPY).
56. Drummuddioch may contain the word *buid(s)each* 'witch, wizzard'.
57. Although other inferences are possible, this apparent agreement between Galloway and Irish place-names in the matter of eclipsis may suggest that the Gaelic settlement of Galloway was later than in other parts of Scotland. For a similar argument, see MacQueen in 'Welsh and Gaelic in Galloway', *Transactions of the Dumfriesshire and Galloway Natural History and Antiquarian Society*, 39 (1953–54), 82.
58. I have noted no convincing examples from Dunbartonshire.
59. Examples from Alexander (1952).
60. Jackson (1972, 163) notes *Aildín* as 'unidentified'. Alexander (1952, 3) identifies it with Aden, Old Deer. Cf. Jackson (1972, 73–4).
61. Examples from Johnston (1904).
62. Examples from MacDonald (1941).
63. Early forms from Watson (1926, 104).
64. Examples from Taylor (1995, 489–90).
65. Only Alexander and Taylor suggest that some instances may represent *n*-stem genitive forms 'with a nominal phrase such as "land of" understood' (Taylor 1995, 42).
66. It is tempting to speculate that the alternation between -*ie* and -*in* originates in the interpretation of the manuscript contraction for -*in*, being a stroke above *i (ī)*, as

either -*í* or -*in*, but the occurrence of -*in* is too widespread to make such a suggestion credible.

67. See Borgstrøm (1940, 178); G. Mac Gill-Fhinnein, *Gàidhlig Uidhist A Deas* (1966), 26. Synchronic evidence provides support for a west-east division of Scottish Gaelic dialects in other respects e.g. preaspiration; use of -*as* future forms; eclipsis and so forth. Nasal stem inflexion occurs only marginally in ScG dialects. See *cù* 'dog', *guala* 'shoulder', *talamh* 'ground', *teanga* 'tongue', *brà* 'quern' in Borgstrøm (1940, 184).
68. See C. Ó Dochartaigh (ed.), *Survey of Gaelic Dialects* (1994).
69. Jackson (1972, 50) argues for an underlying *n*-stem **Bidbe*.
70. Cf. place-names in -*ock* in Scotland which derive from the feminine diminutive suffix -*ag* < -*óg*, e.g. Greenock < Gaelic *Griana(i)g*. See Watson (1926, 201, 447–50). Nieuwenhuis (1985, 87–8) notes that the use of diminutives with familiar objects including 'one's house, home-town or village, country, geographical phenomena' is very widespread. He notes that 'frequent exposure to an object or place creates a special relationship with and an affection for that object or place'. He notes later on, however, that some categories including place-names 'resist diminutive formation on the whole' (Nieuwenhuis 1985, 162) although there are numerous instances in the world's languages of place-names with diminutive forms.
71. For further instances see Hogan (1910), Flanagan (1994).
72. Although the diminutive suffix -*ie*, -*y* is only attested with proper names in Scotland from the fifteenth century, there is no reason to assume that it had not developed at an earlier date. Jespersen (1933, 297) notes that 'pet-names and pet-formations may have existed long in the spoken language without being thought worthy of being committed to writing in an age that was not as apt as our own to record familiar speech'.
73. Cf. *The Scottish National Dictionary*, vol. 5, 249.
74. He compares the raising in 'pity' < Middle English *pite* (Jespersen (1933, 297).
75. Compare the use of the diminutive -*ittus*, -*itta* which is found in late Latin inscriptions only (Marchand 1969, 299).
76. North eastern Scotland, particularly Aberdeenshire, is well-known for its use of the diminutive -*ie*, -*y*. See E. Deith, *A Grammar of the Buchan Dialect* (Aberdeenshire, 1932), 143–4; W.A. Craigie *et al.*, *The Scottish Tongue. A Series of Lectures on the Vernacular Language of Lowland Scotland* (Maryland, 1970), 125–51; Ludovic Kennedy, *In Bed With an Elephant: a Journey through Scotland's Past and Present* (London, 1995), 330–1. I am grateful to my colleague Ronald Black for the latter reference.
77. See M. Lynch, *Scotland: A New History* (London, 1991), 53–73 (esp. 62).
78. He refers to the Dutch loan pinkie (from Dutch *pink(je)*) in Scots. Cf. *The Concise Scots Dictionary* s.v. pink².
79. There is some evidence to suggest a connection between the Gaelic diminutive ending -*ín* /-*é(i)n* and the Scots diminutive -*ie*, although this needs to be investigated further. The Scots word *collie* may derive from Gaelic *cuilein* (< *cuilé(i)n*, *cuileán* or perhaps **cuilín*). Cf. also Scots *lippie* and Scottish Gaelic *lipein*. Ronald Black, in 'Aikey Fair and the Cult of St Féichín', *Scottish Studies* (forthcoming) derives Aikie, the site of the well-known fair in Old Deer, from the saint's name (Mael-) Fhaechín; if correct,

we have yet another instance of Gaelic *-ín* > Scots *-ie*. Cf. *Mal-(F)aechín* which occurs with a stroke over the final syllable (perhaps a length mark) in the Gaelic notes of the Book of Deer.
80. Cf. *Bróccín* in Jackson (1972, 32, VI, 6).
81. Cf. Jackson (1972, 32, VII, 11).
82. Cf. Mal-[F]échí[n] (Jackson 1972, 32, V, 3). The lack of the final *-n* here may be significant in the present context.
83. On the occurrence of these strokes, see Black (1973). See also Ó Maolalaigh (forthcoming).
84. Black (1946, 20) under Alpin suggests this also.
85. On the historical phonology of Anglo-Norman, see Pope (1934, 420–61), Menger (1904, 38–109). For French, see Elcock (1960).
86. However, these forms may represent hypocoristic forms based on truncated forms of the names in question.
87. The representation of Gaelic *-(a)idh* as *-ie, -y* from the thirteenth century is not inconsistent with the development of the dental fricative *dh* in Gaelic, which had lost its dental articulation by the end of the thirteenth century (O'Rahilly 1932, 65).
88. Seen for example in the French word *jardin* 'garden', which derives from Old French *jart + in* (Elcock 1960, 252).
89. The latter may derive from the diminutive form *Goirtín* 'small field'.
90. Other spellings are also attested such as *-ath, -auch, -aycht, -aw, -ay, -a*, etc. e.g. Braiklay (ABD, various) from Gaelic *Breaclach* occurs as *Brauchla* (1531), *Braclay* (1696), *Brechulath* (1236), *Breklaw* (1484), *Brakla* (1595) Alexander (1952, 24); Brucklay (ABD) from Gaelic *Broclach* (?) occurs as *Broklaw* (1489), *Borroklauch* (1530), *Broclaycht* (1573), *Brucklaw* (1732) Alexander (1952, 26); Knapperna from Gaelic *Cnaparnach* occurs as *Knappernay* (1625) etc. Alexander (1952, 79).
91. But note Auchnotroch (LAN), Watson (1926, 201).
92. Nicolaisen (1993, 311) bases his argument erroneously on the development of Old English *hōh* 'heel; projecting ridge of land' which occurs as the final element in place-names like Fogo (BWK), Kelso (ROX), Minto (ROX), Stobo (PEB). These names, he claims, show the loss of the final fricative *h* [χ] from the thirteenth century. The comparison, as envisaged by Nicolaisen, of Gaelic final *-ch* in eastern Scottish place-names with the development of Old English *hōh* in the Border counties is of course invalid. The development of Old English *hōh* in the place-names mentioned by Nicolaisen does, however, provide evidence for the loss of final unstressed velar fricatives in the history of English. It does not provide evidence for the lengthening of [o] to [o:] in eastern Scottish place-names.
93. Assuming, of course that the derivations are correct in each case. For derivations, see *The Concise Scots Dictionary* (1987), ed. Mairi Robinson s.v. *kyloe, clarsach, gluntoch* respectively. The editors tentatively suggest Gaelic *glùn dubh* 'black knee' as the origin of *gluntoch/gluntow*. However, it is difficult to reconcile *-och* with Gaelic *-ubh* unless of course we take this as evidence that *-och* was used in this word to represent a final vowel sound /o/. I would like to suggest *glù(i)nt(e)ach* as a plausible derivation for both forms *gluntoch/gluntow*. Further research of Gaelic elements in the Scots lexicon may well unearth more examples.

94. See historical sections of E. Dieth, *A Grammar of the Buchan Dialect (Aberdeenshire): Descriptive and Historical*, vol. 1: Phonology and Accidence (Cambridge, 1932), 112–13; R. Zai, *The Phonology of the Morebattle Dialect* (Lucerne, 1942), 216–18. Final unstressed velar fricatives seem rare also in other Germanic languages. See R. Jordan, *Handbook of Middle English Grammar: Phonology* (The Hague, 1974), 178–83; E. Prokosch, *A Comparative Germanic Grammar* (Philadelphia, 1939), 82–4.
95. A similar strategy appears to have been adopted in certain Irish dialects as a reaction against the reduction of voiced velar fricatives where final *-gh* (< *-dh*) was retained as *-g* in certain contexts. See O'Rahilly (1932, 71).
96. *The Concise Scots Dictionary* suggests that *cuddoch* is a reduced form of *colpindach*. However, *cuddoch* is more plausibly derived from Gaelic *cullach*, see E. Dwelly, *The Illustrated Gaelic–English Dictionary* s.v. *cullach*.

2

The Uses of Place-names and Scottish History – Pointers and Pitfalls

G.W.S. Barrow

The usefulness of place-name studies as a tool for the historian came comparatively late in the modern historiography of Scotland. It is true that George Chalmers (1742–1825), John Pinkerton (1758–1826) and William Forbes Skene (1809–92) all appreciated the relevance of philology in establishing a reliable chronology and ethnic analysis of the earliest period of the Scottish past which could realistically be called historical, that is to say the fourth to the seventh century AD. But in their time philological studies of the relevant vernacular languages had scarcely reached their infancy, while it was practically impossible to make any serious use of place-name material before there had been any systematic collection and publication of the earliest forms in which the names of places had been recorded. We can see the beginning of a new era when Cosmo Innes (1798–1874), an exceptionally able record scholar, oversaw the publication of volume I and volume II, parts I and II, of the *Origines Parochiales Scotiae* (1851–55). This mixture of topography and history (with a little archaeology thrown in) was far too ambitiously conceived ever to have been finished. Of the twelve medieval dioceses which the work was planned to cover, only Argyll, Caithness, The Isles, Ross and the greater part of Glasgow had been completed when the undertaking came to a halt. For our purposes the most interesting feature of *OPS* is that all the known forms of the names of parishes are set out in chronological order in the heading for each article, while in the text every reference to a locality or property within the parish is given in its original spelling, with dates where known. This was an immense advance on anything that had gone before, and all subsequent place-name scholars have had reason to regret that Innes and his much underrated assistant John Brichan were not able to deal with the huge residual area of Scotland covered by the dioceses of Orkney, Moray, Aberdeen, Brechin,

Dunkeld, Dunblane, Galloway and, above all, St Andrews. Moreover, informative though it was, *OPS* was not a work of toponymic science. From the 1850s until the period between the two world wars most Scottish historians paid little or no attention to place-names as an aid to historical understanding. Curiously enough, the outstanding exception was an amateur in both fields of study, Sir Herbert Maxwell of Monreith, Baronet, who was certainly aware that the place-names of Galloway and the south-west could throw light at least on the early period of Scottish history. In the mean time experts in the field of Norse language studies – mainly Scandinavians themselves, although including several scholars from the Northern Isles and elsewhere in Scotland – had pioneered the difficult historical problems raised by the Viking conquests and settlements of the islands off Scotland's north and west coasts and of sizeable areas of the mainland. The sagas had been largely (and perhaps somewhat over-enthusiastically) published in the nineteenth century. They told a story which seemed increasingly at variance with the results of place-name studies and with what could be learned from the steady advance in archaeological investigation. It is probably no exaggeration to say that this Scandinavian contribution to Scottish historiography decisively established the necessity of place-name study for all historians working in the period before about 1500.

Here, within the scope of a brief chapter, I can do no more than indicate some of the ways in which place-names can act as pointers towards building up a convincing background for important political and ethnic developments and can also confront us with pitfalls which may all too easily obstruct a true reading of the evidence.

Historians still argue vigorously as to whether the Pictish period came to an abrupt and dramatic end in the ninth century or simply faded away gradually, giving place to a Scoto–Pictish amalgam (with, increasingly, a strong input from Scandinavia, especially in the far north and west). In this context the place-name element most often used to demonstrate continuity has been the diagnostically 'Pictish' term *pett* (normally reflected as *Pit-* in modern names). Kenneth Jackson's famous distribution map of *pett-* names, first published in 1955 and many times reprinted, has been for all students of the period the definitive map of the 'Pictish area' of north Britain (Jackson 1955, 147). Jackson also compiled and published in 1955 another map of certain P-Celtic place-name elements other than *Pit-* (e.g. *carden, lanerc, pert, pevr, aber*), but unfortunately his map confined these elements to the country north of the Antonine Wall, suggesting to the unwary reader that these elements, like *Pit-*, were specifically Pictish (*ibid.*, 150).

While we may accept that *pett* was an essentially Pictish term which was not easy for the Scots to replace when they stepped into Pictish shoes, c.850, we have also to grapple with the fact that the word stuck successfully throughout eastern Scotland, although the term *baile*, perhaps the closest Scottish equivalent, was clearly able to establish itself in exactly the same region, perhaps during the second half of the period c.850 to c.1150. Jackson wrote of 'the intimate and growing contact between the Picts and the Gaelic Scots from the fifth to the ninth century, and of the gradual Gaelic infiltration of Pictland... in which colonisation, intermarriage (with the resulting exchange of names), and conversion to Christianity can be shown to have played a considerable part' (*ibid.*, 151). He was surely right to posit a gradual take-over of Pictland by the Scots. Only gradualness and co-existence could account for the fact that the commonest words in northern Scotland for a large, prominent or conspicuous hill are *beinn* (ben) and *càrn* (cairn), not, as in Ireland, *sliabh*. Equally, only a gradual take-over could have allowed the survival of a substantial number of P-Celtic terms which Pictland shared with the 'larger Cumbria' (Strathclyde and Cumbria as far south as Stainmore WML and YON) and parts of Lothian. Of the five listed by Jackson, we can hardly overlook *aber*, 'burn- or river-mouth', since it forms such a basic element in defining an overall pattern of settlement.

Figure 2.1 presents it in relationship to its Q-Celtic equivalent *inbhear* (inver, inner). Jackson's map failed to show Aberlady and (lost) Aberlessic ELO, Abercarf, now Wiston in upper Clydesdale, Abermelc, now Castlemilk in Annandale (NT265054), and Aberlosk, Eskdalemuir DMF (NT263037). Even so, the comparative absence of both *aber* and *inbhear* from south-western Scotland is remarkable and calls for an explanation – as, of course, does the apparently total absence of *aber* between Dumfriesshire and North Wales!

Figure 2.2 presents an element which almost by definition is likely to be P-Celtic, *neued* (Gaulish *nemeton*) 'sacred grove, sanctuary', even though the Q-Celtic cognate (Gaelic *neimhidh*) must lie behind some of our modern reflexes (e.g. Nevie, Nevay). In Scotland at least, *neued* seems never to have been applied *de novo* to a Christian site, although a majority of the known occurrences refer to places where some sort of Christian dedication or facility for Christian worship was provided at an early date. This fact, indeed, may explain why the names have survived. And whether the Christianising agent was Q-Celtic, as must surely have been the case benorth the Mounth, or P-Celtic, as is likely to have been true at many sites between the Mounth and Solway, the survival of *neued* names strongly confirms the theme of continuity. If I am correct in believing that Newtyle (*Neutil*) and (lost) Newtibber (*Neutobir*) embody the *neued* element, then we have a

2: The Uses of Place-names and Scottish History

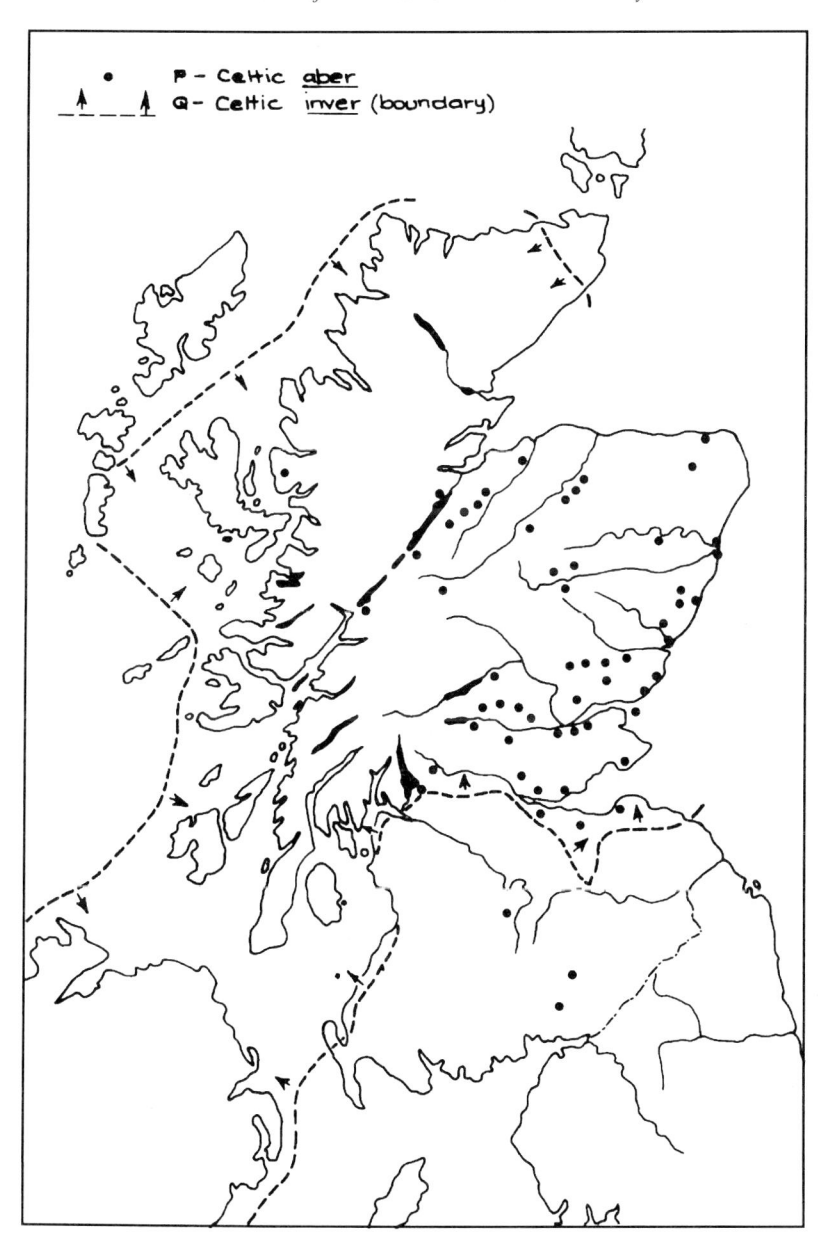

- • P – Celtic <u>aber</u>
- ▲ – – ▲ Q – Celtic <u>inver</u> (boundary)

2.1: Distribution of *aber-* and *inver-* place-names
(Map: J. Renny)

2: The Uses of Place-names and Scottish History

2.2: Distribution of *nemeton*- place-names, with other
localities associated with pagan worship
(Map: J. Renny)

fascinating cluster of names in and under the Sidlaw Hills on the Perthshire–Angus border – Nevay, Newtibber and Newtyle – which suggest that pre-Ninianic, pre-Columban paganism, evidently related to the paganism of Gaul, survived stubbornly enough to have left at least toponymic traces long after Q-Celtic speech had supplanted P-Celtic in this corner of the country.

An element to which only the great Swedish place-name scholar Eilert Ekwall seems to have done justice (although Sir Herbert Maxwell recognised its meaning and importance) is shown in Figures 2.3 and 2.4. This is the P-Celtic word *pol*, which must be distinguished, at any rate as a place-name element, from Q-Celtic *poll* and Old English *pōl*, both of which have the roughly similar meaning of 'pool', 'hole', 'cavity', etc. Maxwell (from his deep knowledge of Galloway and Nithsdale) and Ekwall, who kept an eye open for Scottish examples and parallels, knew that *pol* must have been a British word meaning 'stream', 'flowing water'. It has, remarkably, survived in modern Scots, in the form 'pow' (not to be confused with the word for '[top of the] head', cf. 'poll'), and now tends to have the sense of 'sluggish stream', often referring to a burn which moves slowly through peaty ground or tidal salt-marshes. Ekwall dealt with most of the extant examples in Westmorland, Cumberland and Northumberland. Figure 2.3 shows the general distribution, illustrating a northern British 'province' for this word which corresponds approximately with the *aber* 'province' of Figure 2.2 and the *minit, munet*, etc. 'province' of Figure 2.7. The survival of *pol* well to the north of the Tay, indeed probably as far north as the Moray Firth, reinforces the argument for continuity of settlement and habitation through the period when Q-Celtic speech succeeded P-Celtic. In south-west Scotland, between the Clyde and the Ayrshire coast and especially in upper Nithsdale, the term *pol* for a burn (by no means confined to sluggish watercourses) occurs so frequently that we are forced to conclude that in this region it was the standard word for a small or medium-sized stream, so well established that it survived the appearance of Old English, Gaelic and Older Scots (represented by square on Fig. 2.3).

It is probably unnecessary to labour the argument for continuity when the evidence is so compelling. But the widespread practice of transhumance throughout northern Britain (as, of course, in Wales) is illustrated in Scotland by comparatively abundant place-name evidence which places beyond any doubt the gradual adoption by successive layers of settlers of relevant place-nomenclature. This process has left in southern and eastern Scotland a remarkably solid P-Celtic base, upon which equivalent terms in the languages of later arrivals were planted whenever and wherever this seemed necessary. Figures 2.5–2.7 show some of the elements composing this base.

2: The Uses of Place-names and Scottish History

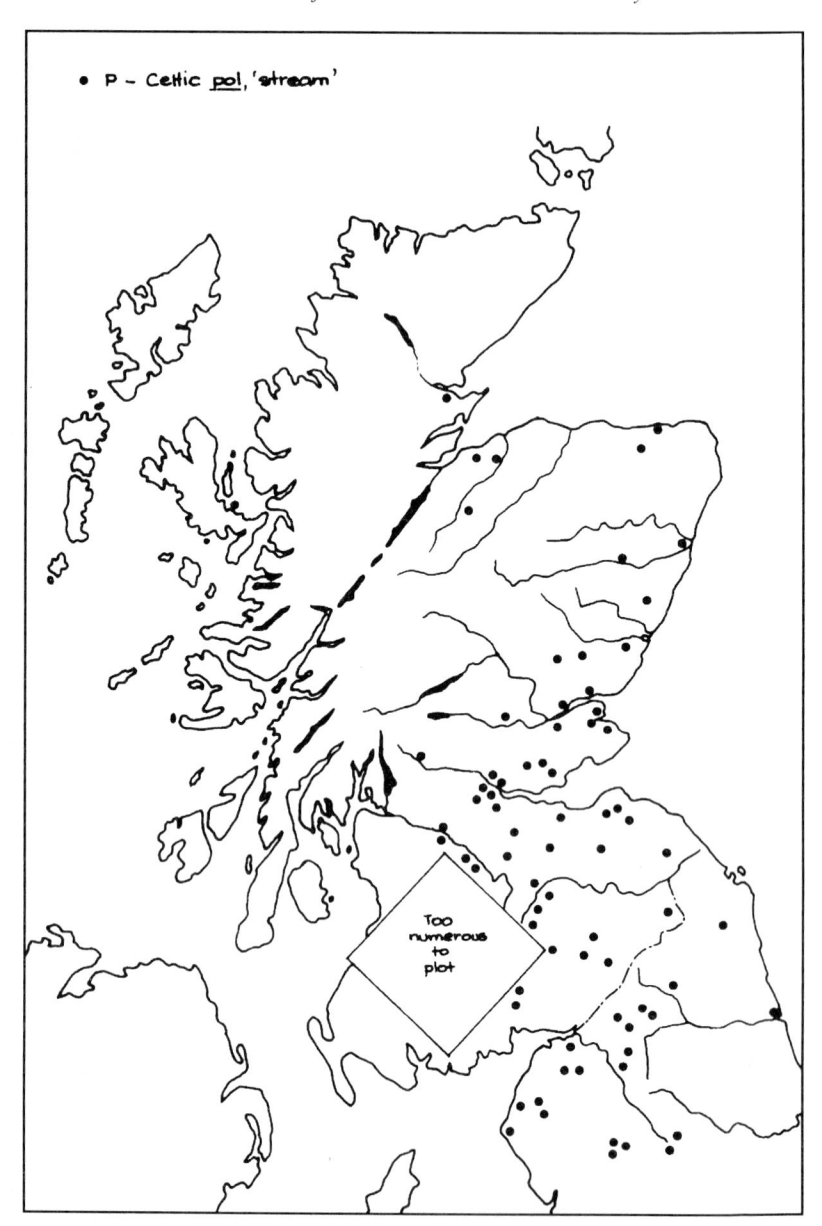

- P – Celtic <u>pol</u>, 'stream'

2.3: Distribution of *pol-* place-names
(Map: J. Renny)

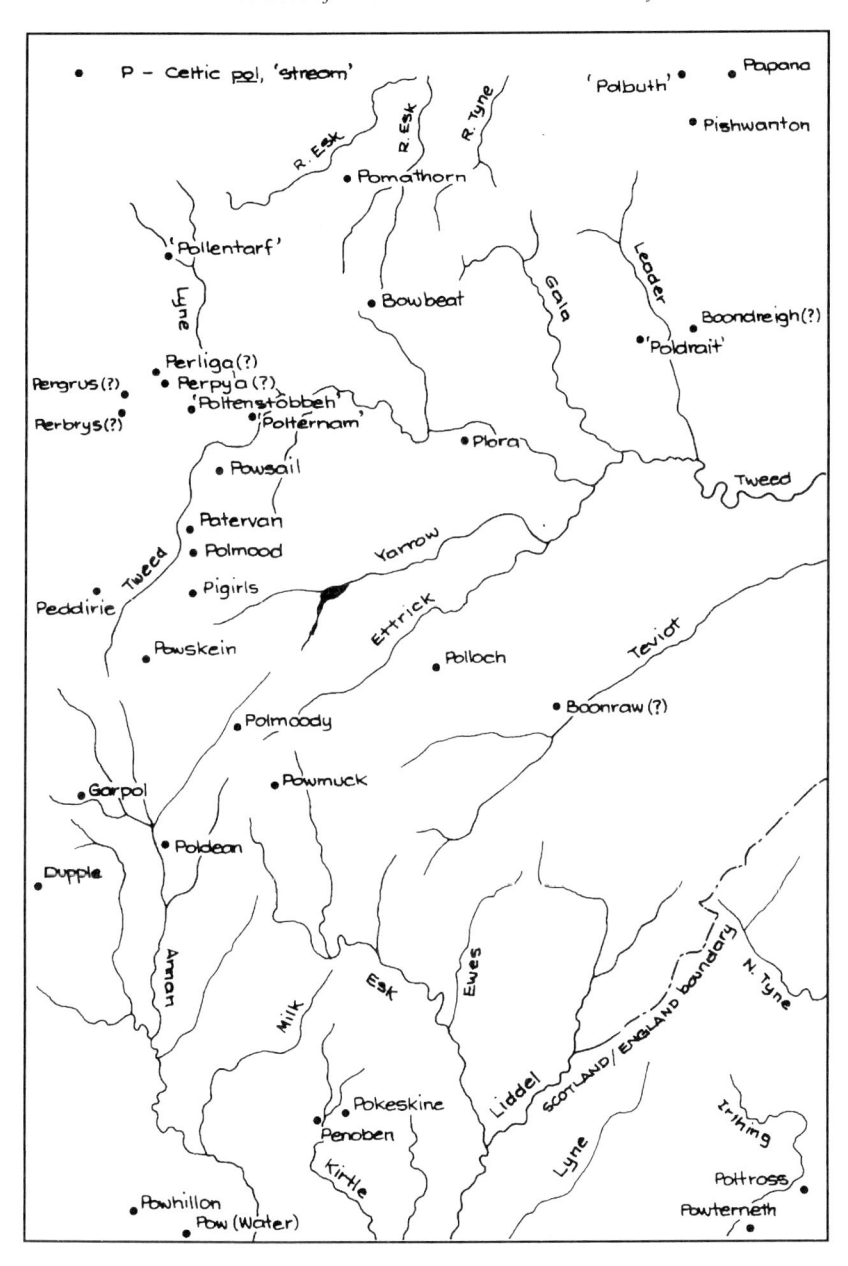

2: The Uses of Place-names and Scottish History

2.4: Central Scottish Borders, showing *pol*- place-names
(Map: J. Renny)

In Figure 2.5 we see the distribution of the only three Scottish survivals of what is familiar in Wales as *ucheldref*, literally 'high settlement', roughly equivalent to the even commoner *hafod* or *hafoty*, 'summer dwelling'. In Scotland the name has come through to modern times as Ochiltree, a name which does not seem to be known north of Forth. We do, however, have Ogilface (*uchel faes*, 'high field') in West Lothian and north of Forth two Ogilvies (*uchelfa*, 'high plain'), one in Blackford parish PER, the other in Glamis parish ANG. In these names the first part corresponds to the Welsh *uchel*, 'high', and there seems little doubt that they are northern equivalents of Ochiltree, albeit with the habitative element replaced by terms of physical geography. Figure 2.6 shows the distribution of the Scottish reflexes of Welsh *pebyll*, 'tent', 'temporary shelter', hence 'shieling'. Unlike Ochiltree *pebyll* or its Pictish equivalent is found north of Forth and Tay; indeed, borrowed into Gaelic it turns up as far north as Abriachan west of Loch Ness (Achpopuli).

The most important of these P-Celtic elements involving the practice of transhumance, the distribution of which demonstrates conclusively that Picts shared with Cumbrians and *Votadini* the custom of exploiting upland grazing in summer, and that incoming Scots took over this custom with the geographical context absorbed rather than replaced, is the term, variously spelled in early sources, which is represented by Welsh *mynydd* (*minit*, *munet*, *moneth*, etc., derived from a supposed British *monijo).

In Wales *mynydd* is used in two distinct but related senses. It either denotes (1) a prominent hill or extensive area of hill ground, a range of hills – examples are Mynydd Mawr (Mynyddfawr) in Caernarvonshire, at SH540547; Y Mynydd Du ('the black mountain') in Dyfed; and Mynydd Du ('Black Mountains') in Powys; or it indicates (2) a stretch of rough pasture, summer grazing which might be on a hill but might simply consist of moorland, not necessarily on an elevated site. Among the extremely numerous instances (especially in southern Wales) we may note Mynydd Llangorse (SO157261), attached to Llangorse, Mynydd Llangatwg (SO180140), attached to Llangattock, Mynydd Castlebythe (SN028297), attached to Castlebythe, and Ysgubor Mountain (SM966304), attached to Skyber Farm. Localities sited on the edge of such areas of rough grazing might be called Penmynydd (Penfynydd), of which I have collected some ten examples drawn from almost all parts of Wales. These were located simply by scanning the 1 inch and 1:50,000 O.S. maps, and no doubt more examples would come to light in a systematic search.

Precisely the same duality is found in Scotland (Fig. 2.7). Here the equivalent of *mynydd* (*minit*, *munet*, *moneth*, etc.) is used for big hills and

2: The Uses of Place-names and Scottish History

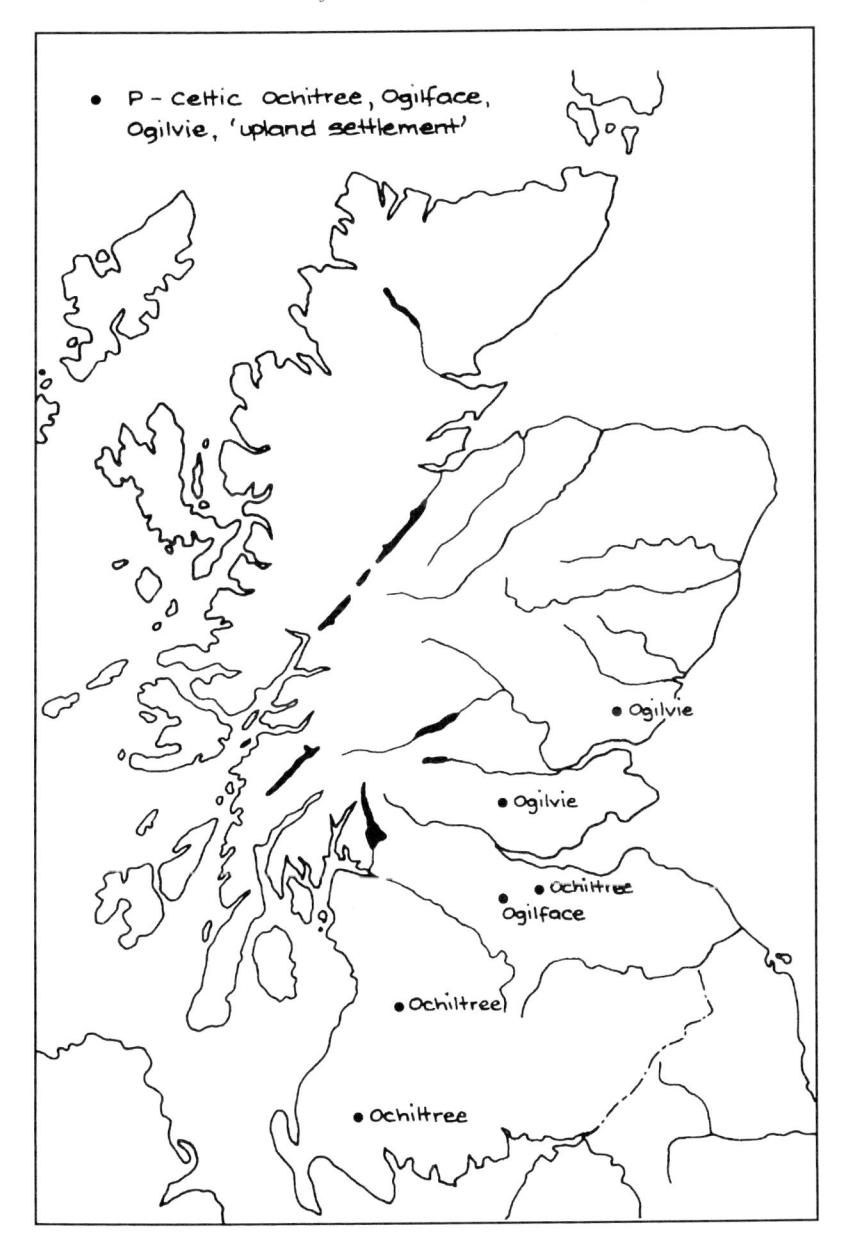

- P-Celtic Ochiltree, Ogilface, Ogilvie, 'upland settlement'

2.5: Distribution of Ochiltree, Ogilface and Ogilvie
(Map: J. Renny)

2: The Uses of Place-names and Scottish History

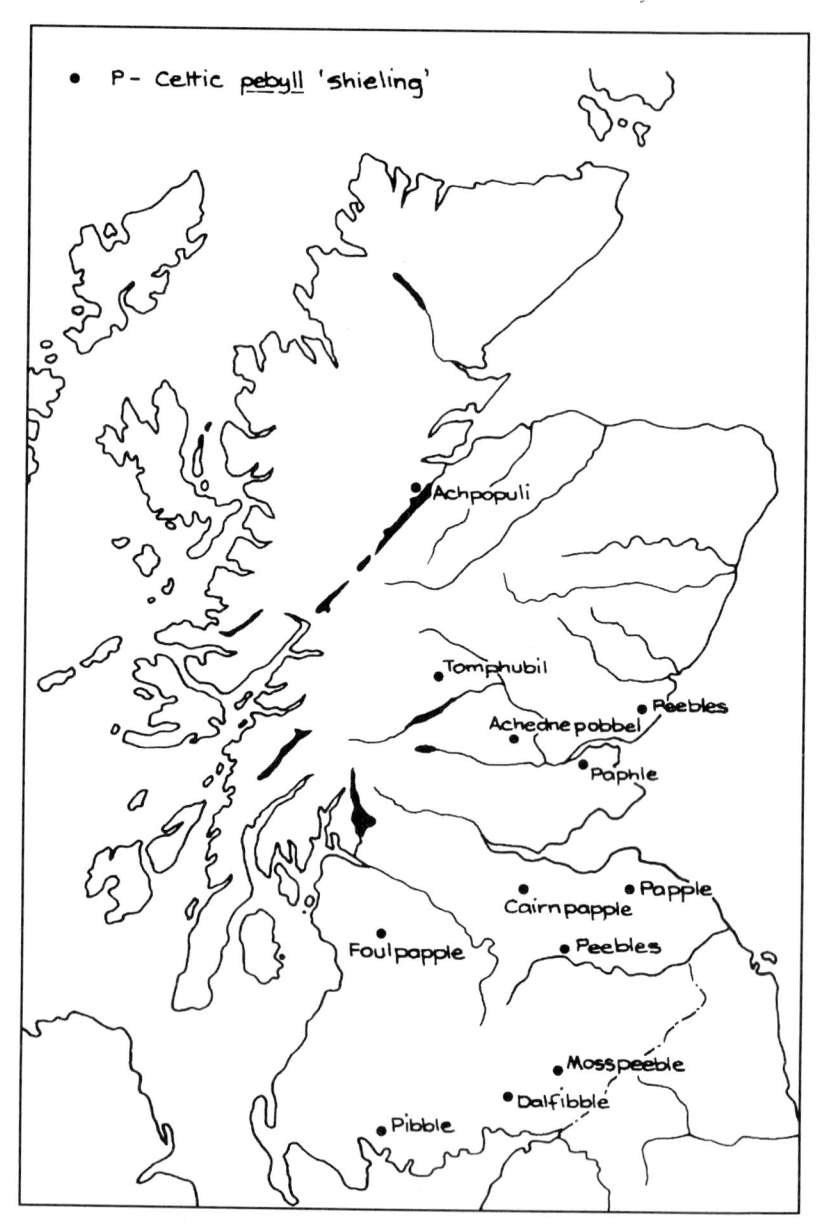

- P- Celtic *pebyll* 'shieling'

- Achpopuli
- Tomphubil
- Achednepobbel
- Peebles
- Paphle
- Cairnpapple
- Papple
- Foulpapple
- Peebles
- Mosspeeble
- Dalfibble
- Pibble

2.6: Distribution of *pebyll*- place-names
(Map: J. Renny)

mountain ranges, as Monadh Liath ('grey hill range'), Monadh Ruadh ('red hill range') and most famously the Mounth itself, the big lump of mountainous territory which stretches from west to east across the middle of Scotland, from Fort William to Stonehaven. The biggish hill known as Monadh Mòr (NN938942) might be seen as a Scottish counterpart of the Snowdonian Mynyddfawr already noted. But the term also (and much more commonly, as in Wales) enters into the formation of names for stretches of rough grazing, as in Mindrum (north Northumberland), Mountlothian (Midlothian), Mendick Hill (Peeblesshire), all south of Forth, and (lost) 'Scleofgarmonth' (Fife, near Loch Gelly, Auchterderran),[1] The Mount (also Fife), Mundie (Perthshire) and north of Tay very numerous examples. Just as in Wales we find Penmynydd (Penfynydd) names situated at the edge of rough grazing so in Scotland we find (with first element 'gaelicised') Kinmount (Dumfriesshire), Kinmonth (Perthshire) and further north a host of names with such forms as Kinmundy, Kilmundie, Kinminity, etc. etc.

Place-nomenclature shows us the linguistic and to some extent the geographical context of transhumance, with continuity indicated at least through the period in which P-Celtic speech gave way to Q-Celtic in south-west Scotland and in the east from the Firth of Forth to the Moray Firth. It is of equal interest that the continuity of transhumance – whether in the fullest sense when it involved small-scale human migrations twice a year from low ground to upland sites and then the reverse, or in the partial sense of shifting beasts from low-lying pasture to upland grazing for the summer months – is strongly reflected in place-names which bring the story forward to the seventeenth or even eighteenth centuries.

It is my contention that the word 'hill' (OE *hyll*) in Scottish usage was very widely used (as, indeed, it still is used even today) to signify 'hill grazing', 'rough grazing' or some such, and not to indicate what Dr Johnson once famously called 'a considerable protuberance'. When today a farmer in upland parts of Scotland says that he is putting his beasts to the hill (or on the hill) he does not mean that they are being forced to bag a few Munros or Corbets. Thus the name Doon Hill (NT685757) does not mean 'a hill called Doon' or (as it might be phrased in modern English) 'Mount Doon', but 'the hill grazing attached to the settlement called Doon', which (nowadays at least) is at NT678760. Just south east of Doon Hill is Pinkerton Hill (Pinkerton), while working westward along the Lammermuir Edge from Doon Hill we have Brunt Hill (The Brunt), Newlands Hill (Newlands), Penshiel Hill (Penshiel), Blegbie Hill (Blegbie), Soutra Hill (Soutra), etc. etc. I would argue that this is a different type of hill name from that found in, e.g. Spott Dod, Deuchrie Dod or Lammer Law in the same locality. Moving

2: The Uses of Place-names and Scottish History

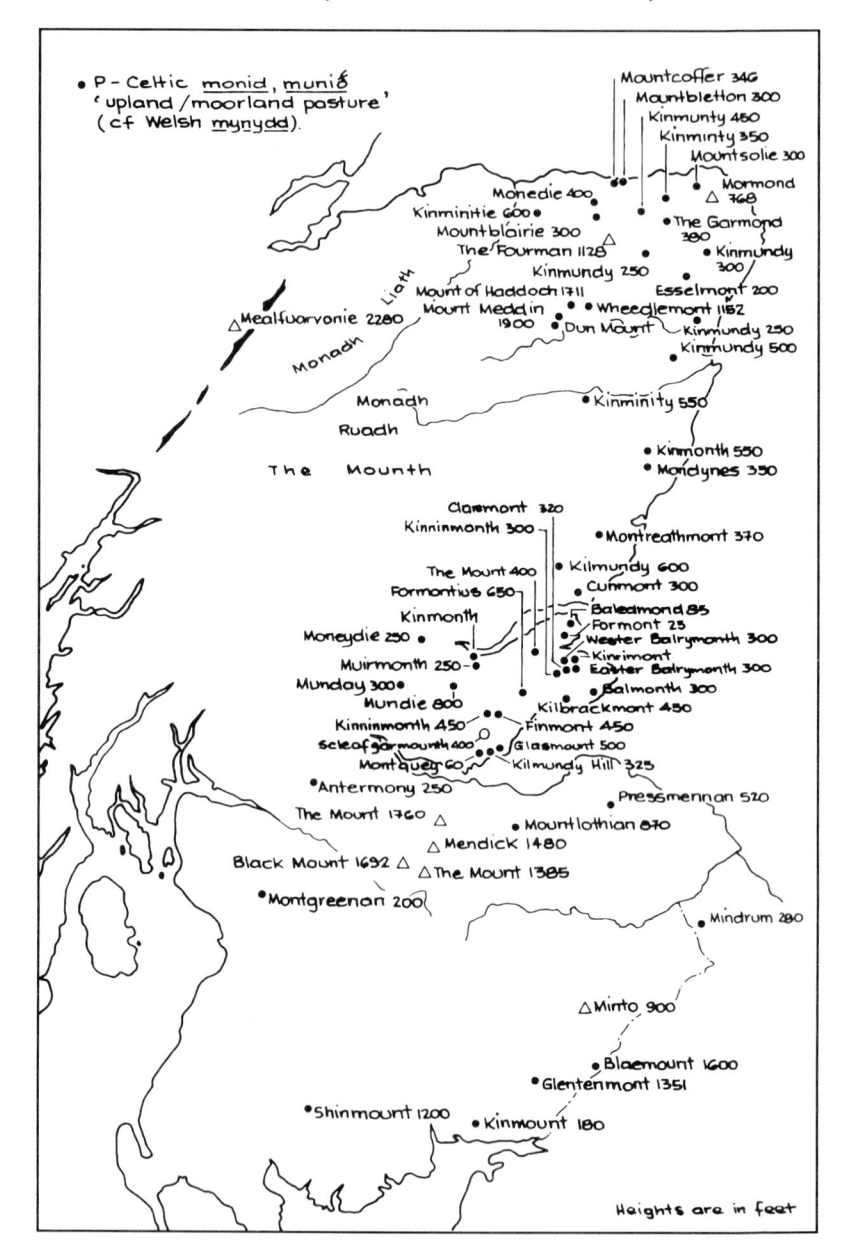

2.7: Place-names containing *monadh*, *munið*
(Map: J. Renny)

2: The Uses of Place-names and Scottish History

north to the Sidlaws in Angus and Perthshire, we may note (working from NE to SW) Dunnichen Hill, NO509497 (Dunnichen), Lownie Hill (Lownie), Hill of Lour (Lour), Hayston Hill (Hayston), Ingliston Hill (Ingliston), Kinpurney Hill (Kinpurney), Henderston Hill (Henderston), Auchterhouse Hill, NO355398 (Auchterhouse), Hatton Hill (Hatton), Newtyle Hill (Newtyle), Hill of Keillor and Keillor Hill (Keillor and High Keillor) and so on and so on. The contrast here would be with such names as Craigowl, NO378400, King's Seat, NO231330 or Dunsinnan Hill, NO214317, where the first two are actual 'hill names' while the third is called after the big concentric vitrified fort which crowns the summit. Across Strathmore from the Sidlaw Hills, in an area where some use of the Gaelic language may be presumed as late as the end of the eighteenth century or even in the early nineteenth, we find the same phenomenon but usually with the word-order showing Gaelic influence. Thus we have Kinclune Hill, NO315568 (Kinclune), Hill of Loyal, NO255503 (Loyal), Hill of Alyth (Alyth), Hill of St Fink (St Fink), Hill of Drimmie, NO185503 (Drimmie), Newtyle Hill, NO050422 (Newtyle, NO049409).

The same exercise could be carried out in many other regions of Scotland in which Older Scots came into use while rural place-names were still in process of formation. The concept of low-lying or sheltered permanent settlement supplemented by high-lying shielings occupied from May to September was firmly established across highland-zone Britain from some very early date until the late-medieval or (in some regions) the early modern period. A comparatively rich stratification of place-names in Scotland provides useful pointers to the concept and practice of transhumance, and to their continuity over at least a millennium.

A relatively unexplored development of Scottish history, for which place-names can certainly serve as pointers even although they are in some respects enigmatic pointers, is the penetration of parts of southern Scotland by the earliest Old English speakers. The received wisdom is that groups or bands of Angles (some coming directly from Lindsey) began to settle the Northumberland littoral about 550 AD and were in a position to expand westward in strength after the battle of Degsastān in 603. Edinburgh, presumably with its hinterland, was in their hands by c.640. An English-controlled Dunbar played a part in the early life of St Wilfrid (born 634), while the English-speaking Cuthbert, presumably an Angle himself, was born about 635 in or on the edge of Lammermuir and within the catchment area of the Scottish monastery at Old Melrose.

If the terms *burg* or *burh* (Fig. 2.8) and *bopl* or *botl* (Fig. 2.9) are rightly to be seen as early elements in Anglian place-name formation their distribution across southern Scotland inevitably raises questions. Both elements turn up further west than we should expect merely from the narrative sources of early Scottish history. In particular, Turnberry in Carrick (*þorne byrig*, 'at the old fortification with thorn bushes'?) and Maybole (*maege boþl*, 'lord's hall of the maidens'?) and 'Neubotle' in Cunningham (*niwe boþl*, 'new lord's hall') are perhaps to be grouped with Prestwick (*preost* or *preosta wic*, 'priest's (priests') farm') and Eaglesham (egles-ham, 'homestead at pre-Anglian Christian church') as indicators of an upper-class Anglian occupation of Ayrshire and East Renfrewshire which leap-frogged across Clydesdale and upper Nithsdale and ought perhaps to be related to the mid-eighth-century acquisition of the 'plain of Kyle' by Edbert, king of Northumbria.

On the east side of the country the spread of both elements seems quite extensive. We note that *boþl* just managed to cross the Forth, in the hybrid name *Ballebotlia*, now Babbet in Kingsbarns parish, East Fife. Perhaps this may be linked with the occurrence of such an early-seeming Anglian name as Pusk (*Pureswic*) in Leuchars parish. And linguistic and cultural inter-communication between Gaelic-speaking Scots and English-speaking Angles in eastern Scotland seems to be illustrated by twelfth-century evidence. For example, an East Fife priest of c.1165–83 had the name Gillequdberit, i.e. *gille-Chutbert*, 'servant of (St) Cuthbert', and at some period now unascertainable (but probably fairly early) the Old English *scir*, 'division', 'cut off portion', was adopted benorth Forth.

Finally, a distribution map (Fig. 2.10) of the element derived from Old English *þreapian*, 'to dispute', 'to argue over' (Scots *threep*) provides a pointer to an active process of boundary definition at a period (twelfth to thirteenth century?) when the use of Older Scots was becoming fairly general throughout lowland Scotland. The implication of the ubiquity and versatility of this word is surely that this period saw the creation of numerous estates presumably carved out of older circumscriptions which now went out of use. This ties in with many surviving charters which attempt (sometimes very ineffectually) to define estate boundaries, and with the record of early perambulations.

If properly used place-names can provide pointers to many historical processes, especially in the period before c.1200. Indeed, it is no exaggeration to say that for the history of Scotland before c.1100 place-names constitute the most important single document. Place-names, however, may provide pitfalls for the unwary. I am not thinking here of the absurd or facetious derivations which used to abound in popular (especially local) histories, of the kind which explained the name Barnweill in Craigie,

2: The Uses of Place-names and Scottish History

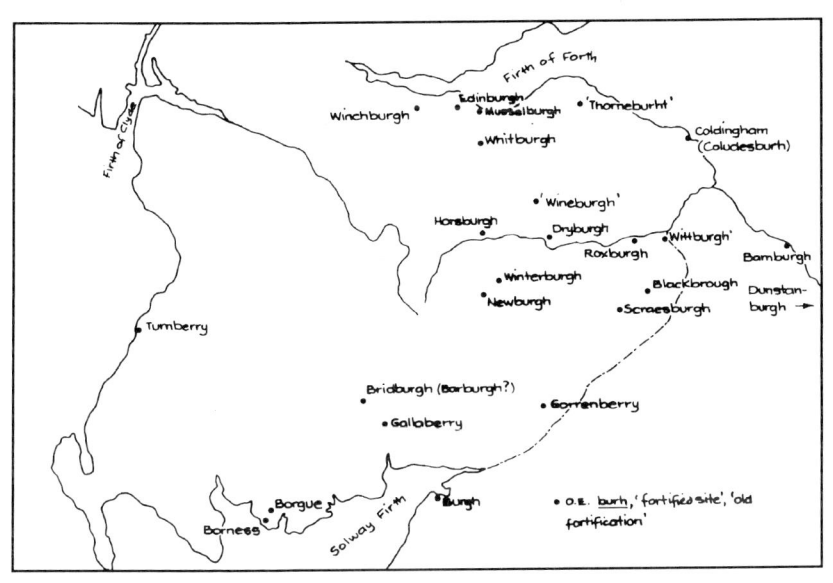

2.8: Place-names containing O.E. *burh/burg* in southern Scotland and northern England
(Map: J. Renny)

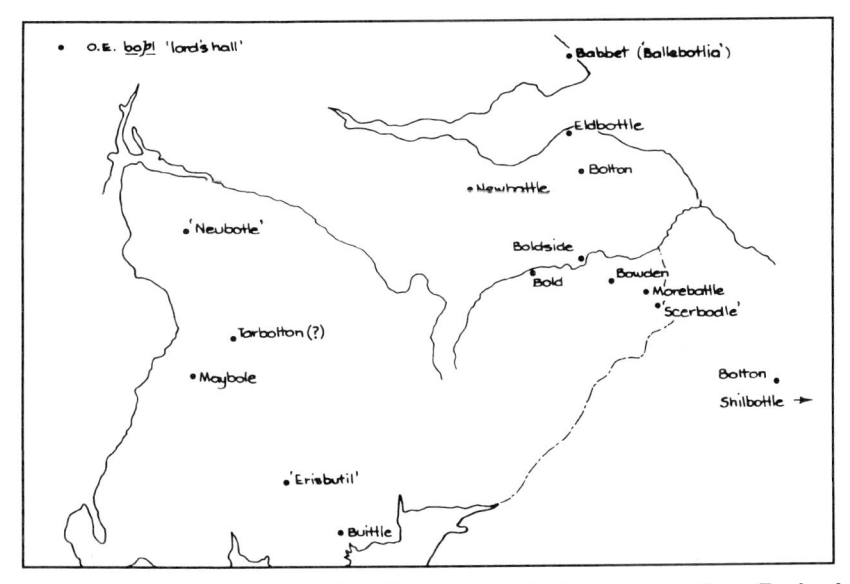

2.9: Place-names containing O.E. *bopl* in southern Scotland and northern England
(Map: J. Renny)

Ayrshire, from William Wallace's setting fire to the barns of Ayr and remarking as his enemies perished in the conflagration 'they burn weel'. Occasionally a puzzling-looking name may in truth be a puzzle. For example, the Fereneze Hills by Barrhead in Renfrewshire owe their odd name to a misunderstanding of the Old French word *fermeson*, 'close season', in charters of the early Stewarts for their abbey of Paisley, which allowed the monks to take hinds but not stags in the Stewarts' forest of Raiss *in fermeson* (when hind hunting was permitted). The odd-looking farm name Clarabad (in Hutton BWK) is derived from Clarembald of Castle Ashby NTP, a twelfth- or thirteenth-century follower of the Olifards, and either he or an ancestor of the same name provides the explanation for Clermiston west of Edinburgh (*Clarembaldestun* > *Clerribaldstoun* > Clermiston).

The pitfalls I have in mind are not connected with popular etymology or occasional oddities. They may lie in wait for us even when the philological analysis and identification are perfectly sound. Problems arise only when we try to use the material to make deductions which have a general validity. A century ago the great English historian F.W. Maitland, discussing the implications which the Danish settlement of eastern England in the ninth and tenth centuries could have had for English freedom in the later middle ages, said 'in truth we must be careful how we use our Dane' (Maitland 1897, 176). We would do well to heed Maitland's appeal for caution when examining the place-name evidence which points to an elusive, almost fugitive, Scandinavian presence in eastern and south-central Scotland at an unknown period between 800 AD and *c.*1050. Where we are dealing with a cluster of place-names of wholly Scandinavian character as is the case with Blegbie, Pogbie, Humbie and *Laysynbi*, the name which lies behind Leaston (all ELO), and probably Begbie less than five miles NE of Leaston, it is surely difficult to interpret this in any other way than by positing a genuinely Scandinavian settlement beneath the Lammermuir edge, lasting long enough for the names bestowed upon physical settlements to endure into a period when a Scandinavian language (or languages) had ceased to be in common use in the locality. But then what are we to make of Smeaton MLO, only eight miles NW of Humbie? Seemingly a simple Old English name, *smeotha tun*, 'smiths' toun', it is in fact recorded early (1154–59) as *Smithebi* (*RRS* i, 183), though found as *Smithetun* in 1150 (*Dunfermline Reg.*, p. 5) and *Smihet[un]* in 1170 (*SHR*, xxx (1951), 45, 49). If this makes it Scandinavian, are we to add Smeaton to our Scandinavian cluster beneath Lammermuir, or do we infer a much more widespread Scandinavian presence in Lothian? Similarly, when we note Corsbie in Legerwood BWK (NT607442) or Newbie in Manor PEB (approx. NT20/33; see Buchan and Paton 1927, 580)

2: The Uses of Place-names and Scottish History

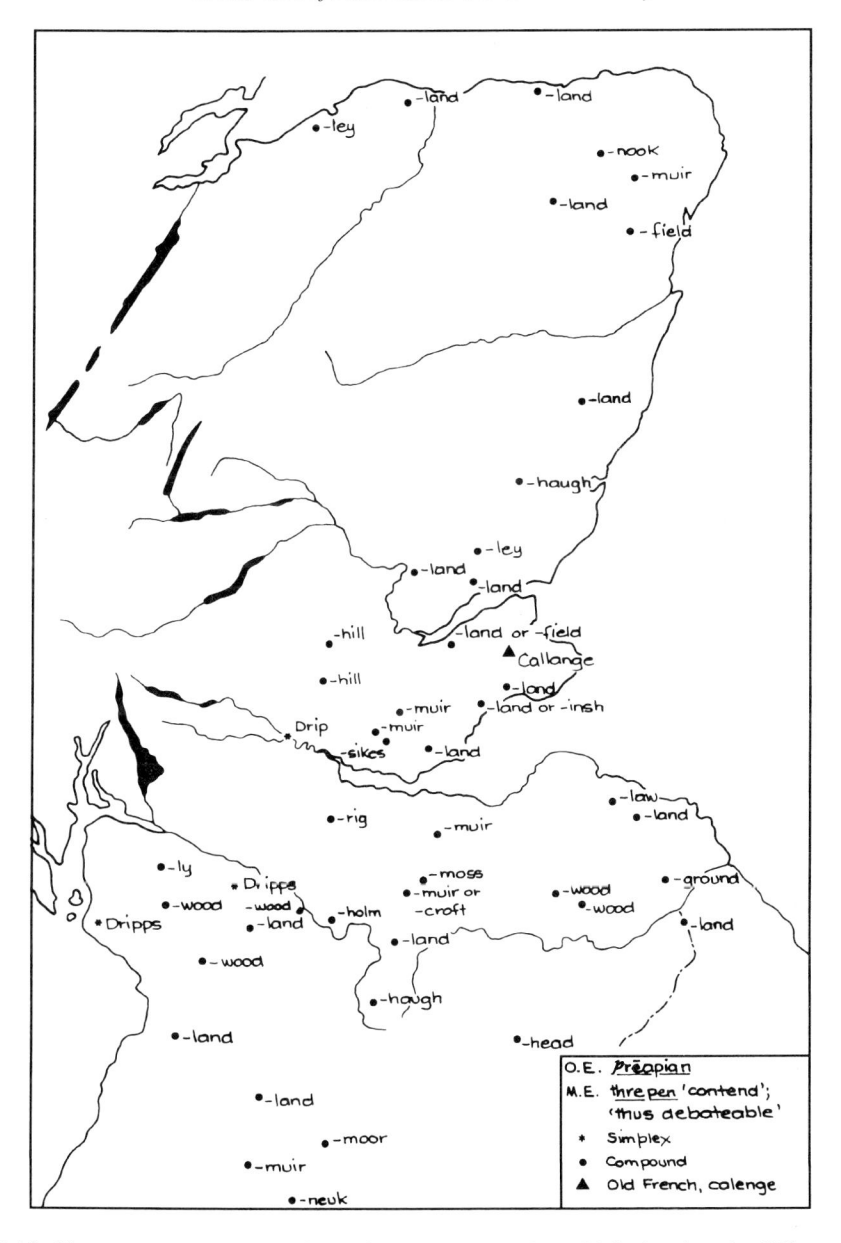

2.10: Place-names containing Older Scots *threep-* 'debateable', showing the different generic elements with which it is combined
(Map: J. Renny)

do we suspect a Scandinavian influx of which there is otherwise no record? Surely not.

We have to make allowance for a considerable borrowing of Scandinavian terms and forms into an Old English place-nomenclature, just as, presumably contemporaneously, we allow for a substantial borrowing into both English-speaking and Gaelic-speaking societies of eastern and southern Scotland of a good many Scandinavian personal names, e.g. Gamal, Liotr, Thor, Thorfinn, Orm, Ulf and Ulfkill, etc. etc.

Among place-name terms which were obviously borrowed in this way I would draw attention to '-bie, -by' (*bý, býr*), 'fell' (*fell, fiall*), 'gill' (*gil*), 'grain' (*grein*) and 'wrae' (*vrá*). Apart from clusters, such as that in Lothian already mentioned and also in south Annandale (Fellows-Jensen 1985) and possibly west Cunningham and north-west Kyle, Ayrshire (e.g. Sorbie, NS245445, Busbie, NS239457, Crosbie, NS216500, Crosbie, NS344302), *bie*-names appear rather fugitively in areas not otherwise known to have been Scandinavianised, e.g. Fife (e.g. Weddersbie, NO261131, Corbie, now Birkhill, NO336234),[2] and Angus (e.g. Ravensby, NO535353). Fell occurs in the Southern Uplands over a wider area than the south Dumfriesshire 'cluster' would indicate, and of course there are the maverick Campsie Fells. Gill is commonly used for a hill burn or ravine in upper Clydesdale, especially Crawford parish. Grain (Scand. *grein*, 'branch') is commonly found in north Dumfriesshire and across into west Roxburghshire, used for small upper tributaries of larger burns or rivers. Wrae ('neuk', 'secluded spot') has a much wider distribution, occurring sporadically up the eastern side of Scotland from Wrae in upper Tweeddale (NT117333) at least as far north as Wreaton west of Aboyne (NJ502993).

Although these terms could not be found quite widely and/or densely distributed throughout several areas of Scotland unless there had been some significant contact with Scandinavian speakers, their use and presence need not, and in many cases clearly does not, imply actual settlement by Scandinavians. They became part of the common vocabulary of inhabitants whose mother-tongues were Old English or Gaelic or possibly even (in a few cases) Cumbric. To adapt Maitland's dictum, we must be careful how we use our Scandinavian place-name terminology.

Place-names arouse the interest of a wide variety of people: philologists, folklorists, geographers, archaeologists, genealogists, people who simply have a love for topography and the human landscape. For the historian place-names contribute one of the tools of the trade. They are seldom of use in providing precise answers to specific questions, but they can be vitally important in establishing that a particular phenomenon or event was a *sine*

2: The Uses of Place-names and Scottish History

qua non behind some other phenomenon or event.

In chronology, it is rare for place-names to provide precise dates but they can be decisive in establishing relative chronologies. As far as Scotland is concerned, place-names are important in setting the scene and establishing general background to events in all periods before *c.*1600. For the historian of settlement in particular, the analysis and evaluation of place-names are absolutely essential. Place-names can tell us a great deal about agrarian patterns, landholding practices, animal and crop husbandry, systems of measurement and assessment of agricultural and fiscal capacity, ecclesiastical specialisms and much else besides. We must, however, follow the rules and be guided by common sense. In general the earlier the recorded form the closer we are to the origin of the name, but this may not always be true. Carstairs LAN first occurs (1170) as *Casteltares*, a form preserved for two and a half centuries or more. Yet the present-day spelling may be a more accurate rendering of the original name, with Cumbric *caer*, 'fort' as the first element, rather than Norman-French *c(h)astel*, 'castle'. But on the other hand the older name for Temple MLO, *Balantrodach*, may illustrate the general rule rather than prove an exception to it. Because *baile nan trodach* means 'stead of the warriors', Watson (1926, 136–7) believed that the name was coined in Gaelic *after* David I had granted the estate to the Knights of the Temple *c.*1140. But the oldest recorded form of the name, dating 1175x1199, is *Plent[r]idoc*, a P-Celtic, Brittonic name of which the first element was surely the Cumbric equivalent of Welsh *blaen*, 'upland' (*Glas. Reg.*, i, p. 37). If the study of place-names has a lesson for the historian it is surely that probabilities are easier to come by than absolute certainties.

Abbreviations

Dunfermline Reg.   *Registrum de Dunfermelyn*, Bannatyne Club, Edinburgh, 1842.
Fraser, *Wemyss*   W. Fraser, *Memorials of the Family of Wemyss of Wemyss*, Edinburgh, 1888.
Glas. Reg.   *Registrum Episcopatus Glasguensis*, Bannatyne and Maitland Clubs, Edinburgh, 1843.
RRS i   *Regesta Regum Scottorum*, vol. i, (*Acts of Malcolm IV*) ed. G.W.S. Barrow, Edinburgh, 1960.
SHR   *Scottish Historical Review*, 1903–28, 1947– .

Bibliography

Buchan, J.W. and Paton, H., *History of Peeblesshire* (1927), vol. iii.
Fellows-Jensen, G., *Scandinavian Settlement Names in the North-West* (Copenhagen, 1985).

2: The Uses of Place-names and Scottish History

Jackson, K.H., 'The Pictish Language' in *The Problem of the Picts*, ed. F.T. Wainwright (1955; reprinted 1980).
Maitland, F.W., *Domesday Book and beyond* (1897; Fontana reprint 1960).
Watson, W.J., *Celtic Place-Names of Scotland* (1926).

Notes

1. I regard this as a good illustration of the tautology which arises through a gradual shift from one language to another, rather than from the abrupt replacement of one language by another. A modern example would be Mount Cairn Gorm, and even Mount Ben Nevis, which is sometimes found in London papers. As I understand it, what has happened is that Q-Celtic settlers have found a name *Garmunt (corresponding to Welsh *garw mynydd*) meaning 'rough moor' or 'rough (hill) pasture'; not *entirely* convinced they have understood *mun(i)t aright, they have prefixed the name with Gaelic *sliabh* 'moor'; the writer of the thirteenth century charter in which this name occurs (Fraser, *Wemyss*, ii, no.4), in a similar quandary, has then added Latin *mora* (from Older Scots *mor*), so that we have, in three languages, 'moor of the moor of rough moor'!
2. For more examples, see S. Taylor, 'The Scandinavians in Fife and Kinross: the onomastic evidence' in *Scandinavian Settlement in Northern Britain*, ed. B.E. Crawford (1995). See also S. Taylor 'Scandinavians in Central Scotland – *bý*- place-names and their context' in *Sagas, Saints and Sacrifice* (provisional title), eds. P.A. Bibire and G. Williams (British Museum Press, 1998 forthcoming).

3

Place-names and Landscape

Margaret Gelling, with illustrations by Ann Cole

When the Council for Name Studies in Great Britain and Ireland held its Easter conference at Aberdeen in March 1984, a highlight of the proceedings was the presentation of the newly published study, *The Place-Names of Upper Deeside*, by Adam Watson and Elizabeth Allen. We were privileged to have lectures from the authors of this beautiful book, and in the discussion which followed several speakers compared the highly informative Gaelic names to the Scots and Scots–English ones in Upper Deeside, to the detriment of the latter. It was felt that Gaelic names with meanings like 'little hillock of the speckled field', 'corrie of the divided pool', 'burn of the sunny stream-branch' were more communicative than Scots and Scots–English names like Dikeheid Cottage or the ubiquitous Milltoun. I said that this was not a fair comparison, because truly comparable non-Gaelic names were not the recently-coined ones in Gaelic-speaking areas, but rather the early Anglo-Saxon names of English-speaking regions. The feeling of the meeting was against me, however, and the failure to make my point still rankles a little. So I am glad to have this opportunity to present to a Scottish audience some evidence of precision and subtlety in English topographical names of the post-Roman period.

The Anglo-Saxons seldom used phrase-constructions in place-name formation. Their commonest place-name types consist of a combination of two nouns or of a noun and a personal name, or of a noun qualified by an adjective. This produces, to modern perceptions, a spurious impression of simplicity and shallowness of meaning, and this apparent simplicity misled some great scholars, most notably Sir Frank Stenton, into dismissing names which translate into modern English as 'fern hill', 'long valley' and so on as having little to offer to students of the Anglo-Saxon period. The key to Anglo-Saxon topographical naming lies, however, in the precise use of words which get blanket translations like 'hill' and 'valley' in the impoverished modern English vocabulary. Under these general headings lie large groups of Old English words, each of which denotes a different type of hill or valley, offering quite different possibilities for settlement-growth and land-exploitation. The Anglo-Saxons had a vast and subtle topographical vocabulary which can be decoded by field-work. This absorbing hobby has taken over my life and that of my geographer colleague, Ann Cole. I can only present a very small sample of our findings here, but I hope to be able to convince you that by going out and using our eyes we have ascertained the precise meanings of a great many place-name generics. In addition to the richness and variety of the generics there is infinite scope for fine tuning in the qualifiers of compound names.

I must stress that the few examples I have included in this chapter represent the rule, not the exceptions. The naming system which has been deciphered by our field-work has been found valid in most of England. It does not work so well in Devon, and this combines with other evidence to indicate that the full glory of the topographical vocabulary belongs to the

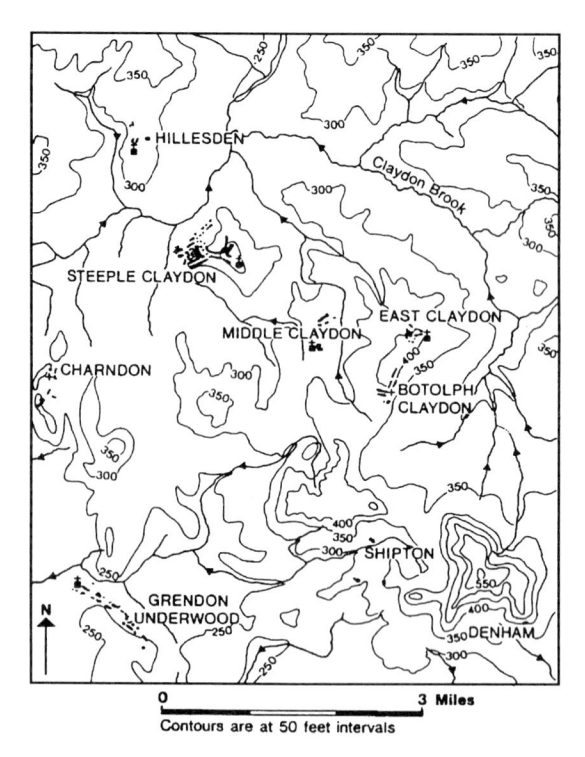

3.1: Buckinghamshire *dūn* country

3: Place-names and Landscape

earliest centuries of Old English speech. I shall not attempt to deal with the historical, linguistic and philosophical questions raised by the fact that with few exceptions this vocabulary is used in the same way from Kent and Dorset to Northumberland and Westmorland, but I hope that my illustrations will convince you that this is so.

The most easily illustrated topographical words are the hill and valley ones, and I have put six of the first and three of the second in the Appendix. The full number of words in these categories is about 40 and 30 respectively.

Dūn means a great deal more than is conveyed by the modern translation 'mountain, hill'. Its commonest use in settlement-names is for low hills which provide good settlement-sites, usually in areas of relatively low relief. On Figures 3.1 and 3.2 you can see how, in the region dominated by the Chiltern Hills, the *dūn* names cluster on the clay plain to the north of the

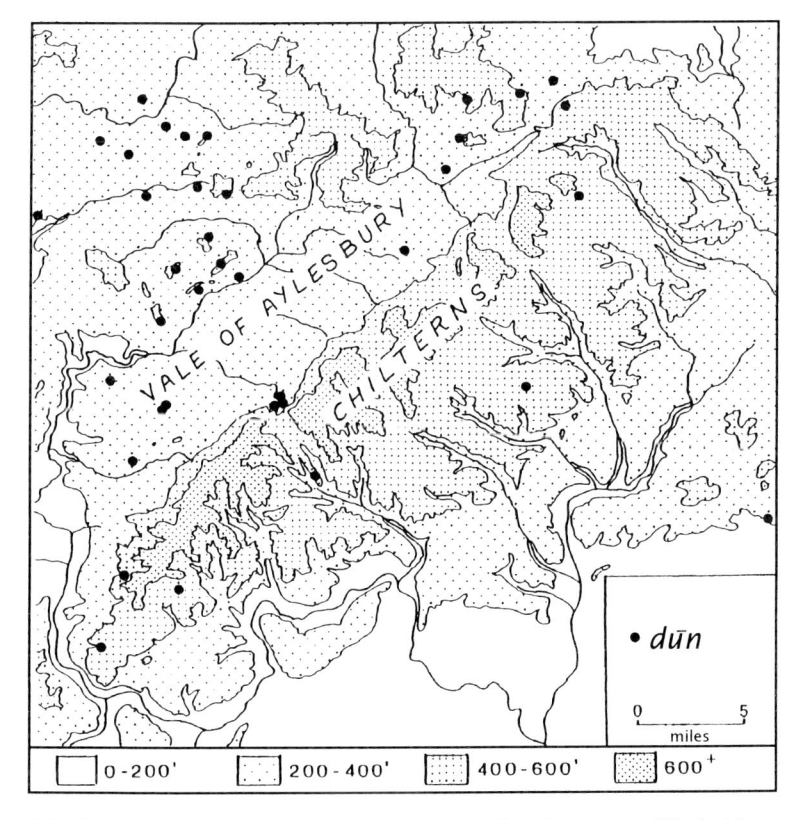

3.2: The Chilterns and the Vale of Aylesbury, showing parts of Berkshire, Buckinghamshire, Oxfordshire, Bedfordshire, Hertfordshire and Middlesex

Hills. The plain stretches from Bedfordshire, through Buckinghamshire to Oxfordshire, and *dūn* is the predominant generic in major settlement-names in this wide area. The contour map (Fig. 3.1) is mainly intended to illustrate the height range into which the majority of *dūn* names fall. They can be lower – Hedon near Hull barely makes 25' (7.6 metres); or higher – Chelmorton in north Derbyshire is at 1250' (381 metres). This map also illustrates the basic, inviolable rule of place-name studies, which is that you must interpret names by early spellings, not by modern forms. Despite the modern form of some of them, all names on the map contain *dūn*.

Figure 3.3 shows four *dūn* sites, chosen from a large collection. Garsington, Billington and Cuddesdon are in the *dūn* country north of the Chilterns. Stottesdon in Shropshire is a valuable example because it has all the characteristics of what geographers call a 'central place' (Fig. 3.3d). The *dūn* villages have English names, but they cannot be English foundations. On sites like these there must have been settlements when the Anglo-Saxons came. What we see here is not (as is frequently asserted) the coining of an Old English hill-name which was subsequently transferred to a settlement. It is the application to ancient settlements of a new English name, the generic of which embraces both the habitations and the site. The element *dūn* in a name would convey to an Anglo-Saxon both the nature of the site and the likelihood of the settlement being a high-status one.

The word *beorg*, also translated as 'mountain, hill', is consistently used in place-names for a hill which has a continuously curving outline. These hills are usually smaller than *dūn*s, and they seldom provide sites for whole villages, though sometimes the church is on the *beorg*. They are frequently occupied by a single farm. In areas of Norse settlement it is often impossible to say whether the English word or the Norse equivalent, *berg*, is involved. It seems necessary to conclude that in northern England the Norse word was used in exactly the same way as the English one.

The illustration of Gran(d)borough speaks for itself (Fig. 3.4a). The next two, Bromsberrow and Edlesborough, make a good pair because it is evident that before the church was built the *beorg* at Edlesborough must have borne a close resemblance to the one at Bromsberrow (Figs. 3.4b and c). Rook Barugh in the North Riding of Yorkshire and Birkland Barrow in Lancashire might look to a casual observer like two views of the same hill, but they are a long way apart, on either side of the Pennines. Scaleber and Brownber provide further examples of the north-country use of the term. Scaleber has a specifically Norse qualifier, and the *berg* in this example is a drumlin.

The name which might have destroyed the concept of Old Norse *berg* being used for rounded hills is Roseberry Topping, in the North Riding of

3: Place-names and Landscape

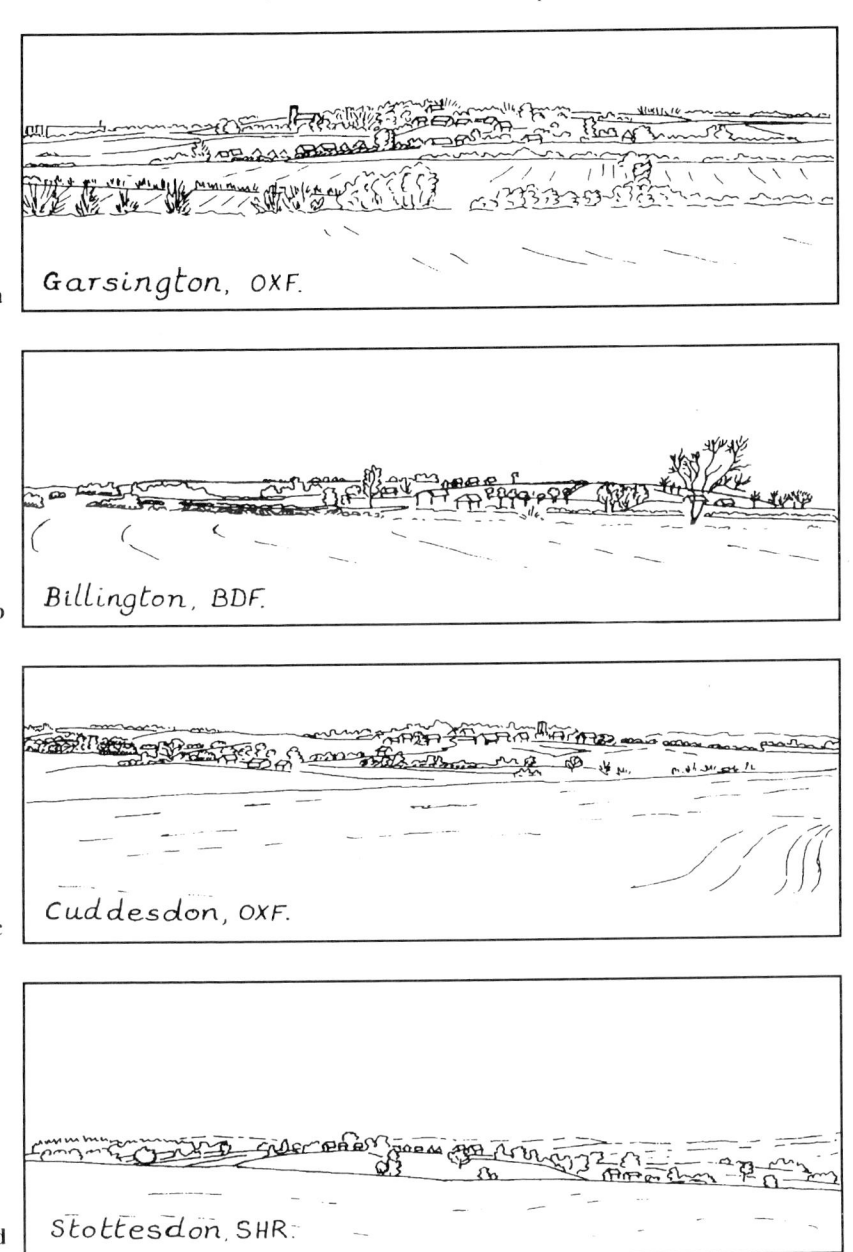

a Garsington, OXF.

b Billington, BDF.

c Cuddesdon, OXF.

d Stottesdon, SHR.

3.3: Illustrations of places with names containing Anglo-Saxon *dūn*

3: Place-names and Landscape

a Granborough, BUC.

b Bromsberrow, GLO.

c Edlesborough, BUC.

d Roseberry Topping, YON.

3.4: Illustrations of places with names containing Anglo-Saxon *beorg* or Old Norse *berg*

Yorkshire (Fig. 3.4d). Roseberry, which means 'Othin's hill', is the only certain reference in England to Old Norse paganism. The main hill, the Topping, is steeply conical, but beside it, as you can see from the illustration, is a good example of a rounded *berg*. Surely this is *Othinsberg* and the other, Roseberry Topping, is the 'peak by *Othinsberg*'. Visible from the site is Langbaurgh, a long hill which gave name to a wapentake.

The Anglo-Saxons had a great many hill-terms, but despite this they borrowed a Welsh one. In Crookbarrow and Creechbarrow (Figs. 3.5a and b) this word has been glossed by *beorg*. The hills are about 100 miles apart, both beside the M5, so they would have been visible from the earlier Roman road. It is my conclusion, based on a large proportion of the hills for which *crūc* is used, that the word was applied to isolated hills which make a particularly striking visual impact. In the cases of Crookbarrow and Creechbarrow the hills were *beorg*-shaped, so the two words together give a precise description. In the case of Crutch (north of Droitwich) the shape is different, but the visual impact is equally striking (Fig. 3.5c). Crouch Hill near Highworth in Wiltshire (Fig. 3.5d) is one of a number of instances in which (despite what was said earlier) the combination of a feature with a Crouch or Crook name justifies an assumption that the Welsh word is involved, although there are no early spellings.

The next three words in the Appendix are used in place-names for hill-spurs: **ofer* and *ōra* denote ridges of the same configuration, which contrasts with that of many of the ridges designated by the anatomical term *hōh*, 'heel'. Ivinghoe and Tysoe (Figs. 3.6a and b) are 'heels', as are Weo (Fig. 3.6c) and Rainow. Landscape features of this shape do not occur in all parts of the country, and where they are absent, as in East Anglia, *hōh* is used more casually for any spur of land. The precise use was current in many areas, however, including Northumberland, where Ingoe is a particularly fine example (Fig. 3.6d).

*Ofer is only known from place-names, though it is related to the recorded word *ōfer*, 'bank, shore, edge, usually of the sea or a river'. In spite of the different vowel-length, the two words may have been perceived by Old English speakers as having somewhat the same meaning. There are, however, a great many *ofer* names which have no relationship to water, and it seems necessary to assume a specialised use for inland features of the term with the short vowel. Occasional early spellings with *u*, as in *Ufre* for Over in Cheshire, establish that there was a word with short *o*. Whatever the semantic difficulties, *ofer* was consistently used all over England for ridges of an instantly recognisable form. They do not rise to a point, and they have convex, as opposed to concave, shoulders (see Fig. 3.7).

a Crookbarrow, WOR.

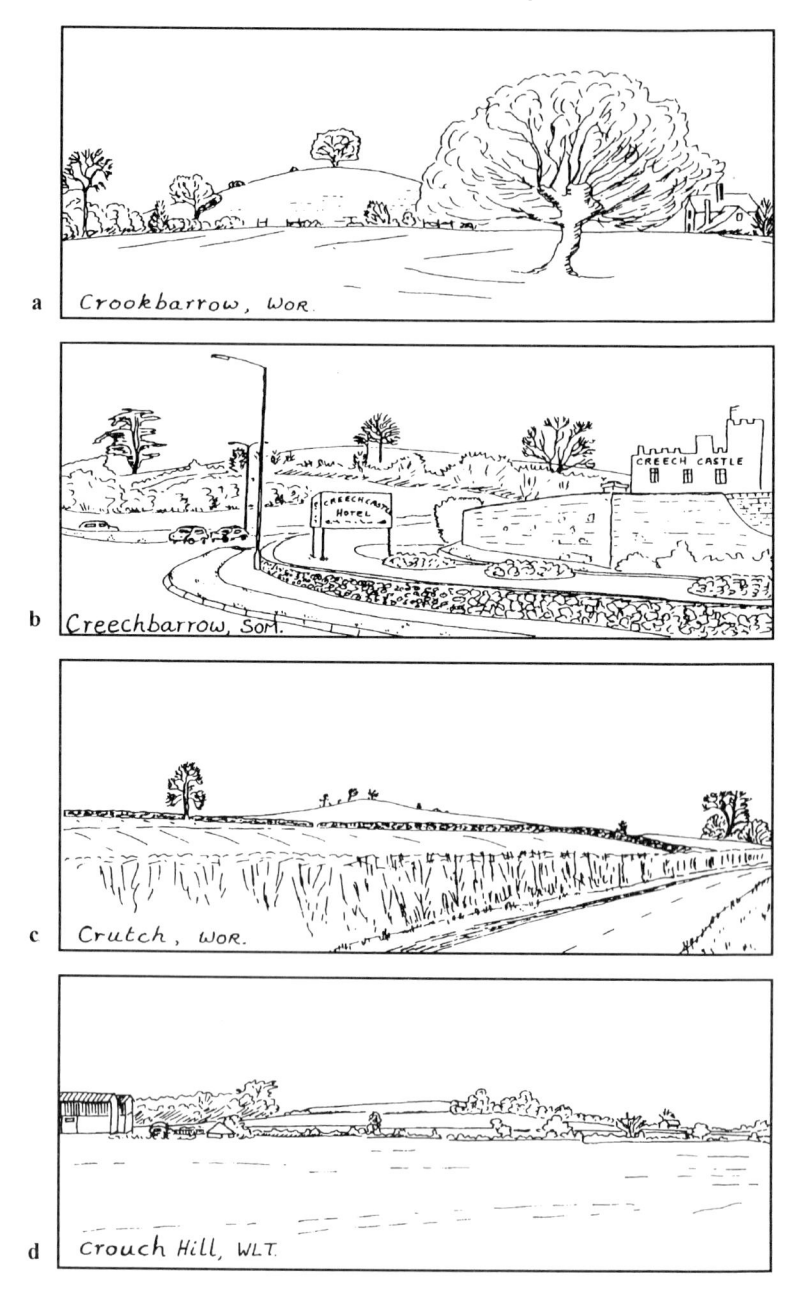

b Creechbarrow, SOM.

c Crutch, WOR.

d Crouch Hill, WLT.

3.5: Illustrations of places with names containing Welsh *crūc*

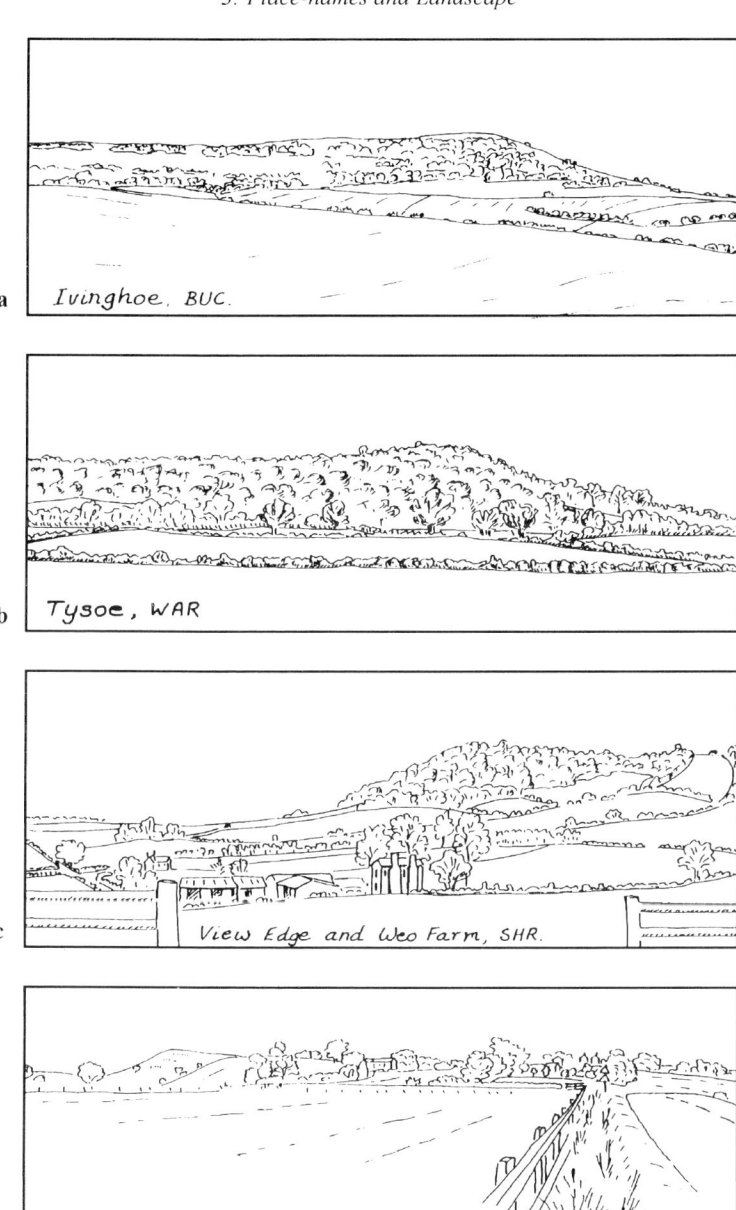

3: Place-names and Landscape

a Ivinghoe, BUC.

b Tysoe, WAR

c View Edge and Weo Farm, SHR.

d Ingoe, NTB.

3.6: Illustrations of places with names containing Anglo-Saxon *hōh*

3: Place-names and Landscape

a Haselor, WAR.

b Shotover, OXF.

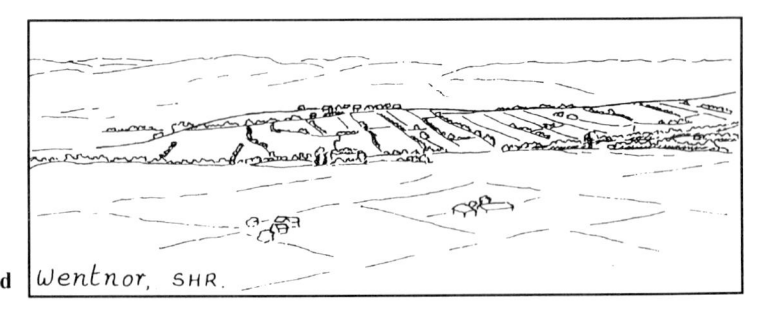

c Wellingore, LIN.

d Wentnor, SHR.

**3.7: Illustrations of places with names containing Anglo-Saxon *ofer*

3: Place-names and Landscape

3.8: Illustration of a place with a name containing Anglo-Saxon *ōra*

The third ridge-term, *ōra*, which is geographically limited to the south of England, has been shown by Ann Cole to refer to land-forms similar to those for which **ofer* was used (Cole 1989 and 1990). The ridges were often bigger, like the great *ōra*s in the Chilterns which have Chinnor (Fig. 3.8) and Lewknor at their foot, but this southern usage is due to the occurrence of larger ridges in the south rather than to any difference in configuration.

Pershore, the furthest north example of an unequivocal *ōra*, is by the River Avon, and *ōra* here has been taken to mean 'river-bank'; but as Pershore is one of a long series of settlements all bearing the same relationship to the river it seems more likely that the name refers to the distinctively shaped ridge on the riverward slope of which the town is situated (see Fig. 3.9).

The two maps (Figs. 3.10 and 3.11) show the distribution of **ofer* and *ōra*, and their relationship to ancient travel routes. The perceptions of travellers may have played a significant part in the evolution and application of this naming system.

Turning now to valley terms, it is instructive to look again at the map of the Chilterns (Fig. 3.12). Here, as elsewhere, the two main terms are *cumb* and *denu*, and settlement-names containing these occur in sharply contrasting situations. The distinction between the long, open *denu* and the short, closed-in *cumb* is a vital one as regards prospects of growth and expansion. Many of the Chiltern settlements in the five- or six-mile-long valleys which characterise the dip-slope have doubled to give Great and Little, or trebled, as in Upper, Middle and Lower Assendon, which lie in the beautiful valley illustrated in Figure 3.13a. This use of *denu* is country-wide except for Devon. In northern and eastern England there is sometimes an interchange in early spellings with Old Norse *dalr*.

3: Place-names and Landscape

3.9: A view of Pershore, from Pensham Hill, from T. Nash's *Collections for the History of Worcestershire*, 1799

3: Place-names and Landscape

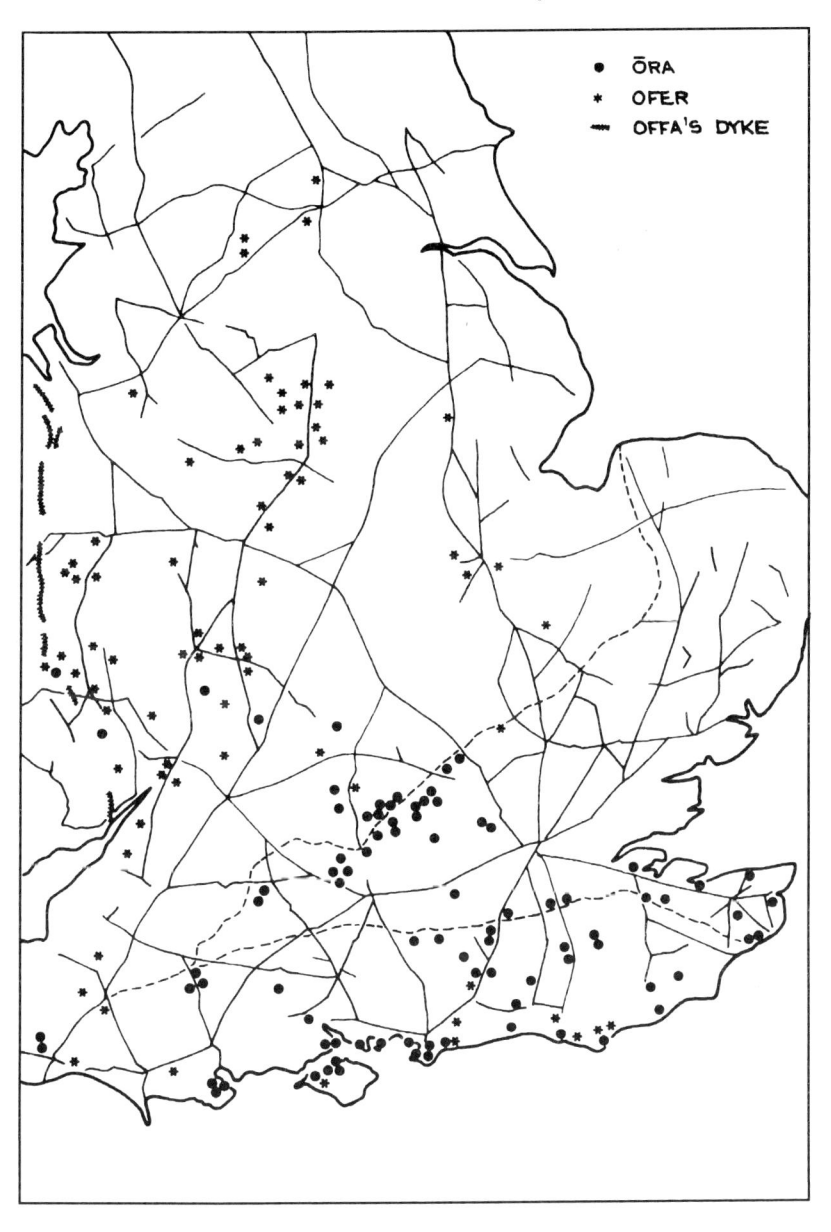

3.10: Distribution of _ōra-_ and _*ofer-_ place-names in England, showing ancient travel routes
(Map: J. Renny)

3: Place-names and Landscape

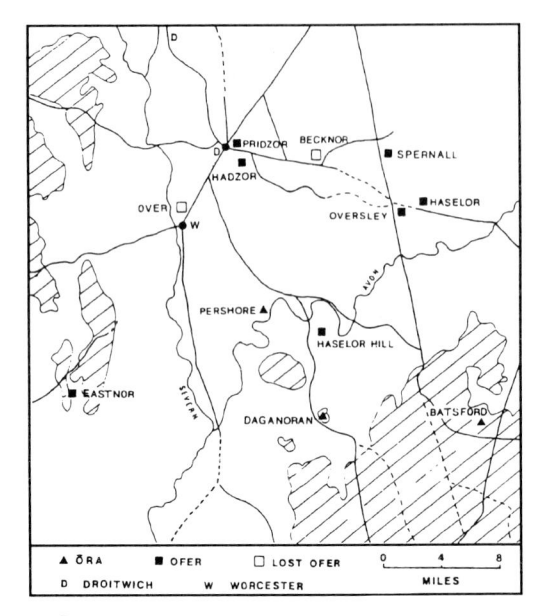

3.11: *Ōra-* and **ofer-* **place-names in the West Midlands, showing the Droitwich Salt Ways**

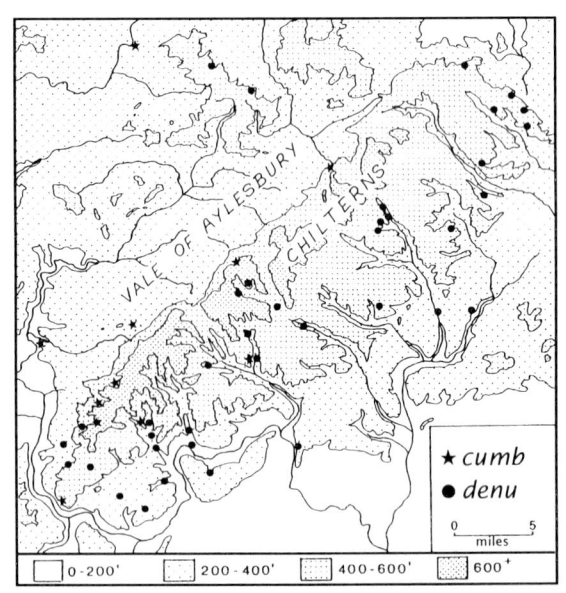

3.12: Relative distribution of *cumb* and *denu* place-names in the Chilterns, south-east England (contours in feet)

3: Place-names and Landscape

A *cumb* settlement, by contrast, had limited scope for expansion and was frequently in a cul-de-sac, so not in contact with the outside world in the same way as a *denu*, which had a through-road running along it (see Fig. 3.14). Valleys of the *cumb* type were so common in the south-west that by the time English speech came to Devon *cumb* had become the general word for 'valley', so it had to be qualified by adjectives, 'wide', 'narrow', 'short', 'long' and so on. Widdecombe in the Moors lies in a wide basin.

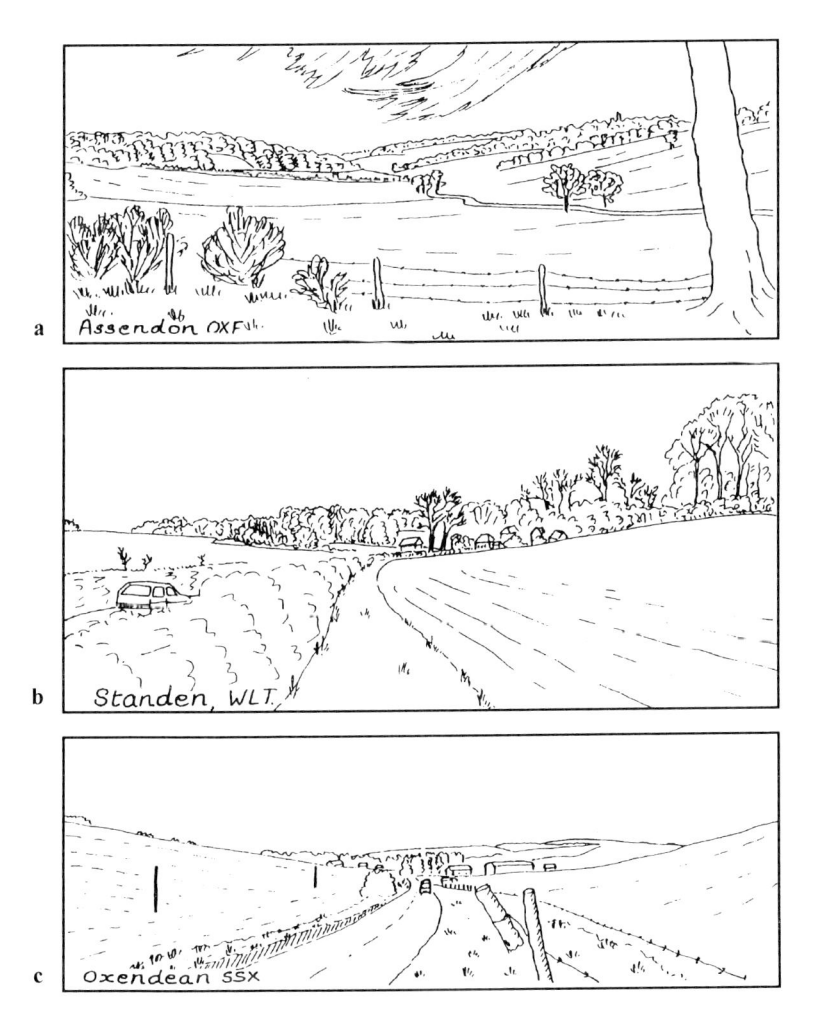

a Assendon OXF

b Standen, WLT

c Oxendean SSX

3.13: Illustrations of places with names containing Anglo-Saxon *denu*

3: Place-names and Landscape

a Encombe, DOR.

b Whitcombe, SOM.

3.14: Illustrations of places with names containing Anglo-Saxon *cumb*

As with hills, the Anglo-Saxons had many words for valleys, but in some parts of England they felt the need of a term which defined something other than size and configuration. In the Welsh Marches, along the edges of the Pennines, and in the Scottish borderlands, there are settlements in valleys whose dominant characteristic is seclusion. This I take to be the significance of *hop*, a word of obscure origin, only recorded in *Beowulf*, and there not used for a valley. With no ascertainable ultimate etymology, one literary occurrence, and a mass of settlement-names, we must surely accept the testimony of the last, and these point to a meaning 'secluded place'. The seclusion is sometimes due to location in marsh or moorland, but much the most frequent use of *hop* is for valleys which do not conform to a single pattern as regards shape, but which all have the special quality of offering a tucked-away settlement-site.

3: Place-names and Landscape

Hope Dale in Shropshire (Figs. 3.15 and 3.16), between the parallel ridges of Wenlock Edge and the Aymestrey Limestone escarpment, contains a series of settlements with names in *hop*, lying in funnel-shaped openings where springs rise. Other items in the concentration of *hop* names in north Herefordshire and south Shropshire are in variously shaped valleys. Figure 3.17 shows Stanhope in Peeblesshire, a typical example from the Scottish Borders, where the word is of frequent occurrence. Although not a feature here, a constricted entry is a common characteristic of *hop* valleys, as at Cowpe in Lancashire.

3.15: Aerial photograph of Hope Dale SHR, looking north

The Anglo-Saxons had a rich vocabulary, also, for wetland features, for what, in my 1984 book, I called 'Marsh, Moor and Flood-plain' (Gelling 1984, 33–61). The most important wetland settlement-term is $\bar{e}g$ 'island', a term which we know to have been used much more frequently before AD 750 than it was in the later part of the Anglo-Saxon period. Ancient settlement-names with $\bar{e}g$ as generic cluster in areas liable to flooding, where a very slightly raised patch of ground could be of vital importance in the siting of dwellings. The islands may be very slight, as at Kingsey in Buckinghamshire, or they can be more pronounced, like the most famous *ēg* of all, Athelney in Somerset (Fig. 3.18a). The Somerset Levels are, of course, one of the areas where *ēg* is frequent, and Middlezoy and Othery (Fig. 3.18b) are neighbouring island settlements there. Important clusters of *ēg* names occur along the middle Thames in Berkshire and Oxfordshire, and in the Fenlands of eastern England.

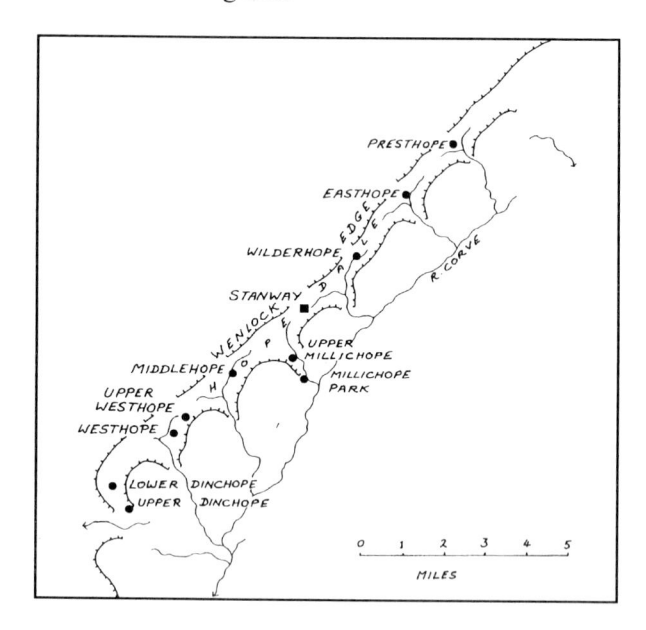

3.16: Map of Hope Dale SHR

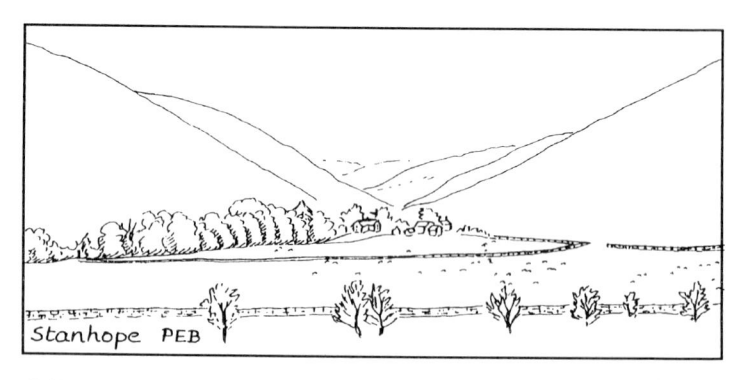

3.17: Illustration of a place with a name containing Anglo-Saxon *hop*

3: Place-names and Landscape

3.18: Illustrations of places with names containing Anglo-Saxon ēg

Wæsse is a term only known from place-names. It is limited to Anglian areas and only occurs in about a dozen names, two-thirds of them in the West Midlands. It is related to words meaning 'wet', and has been assigned the meaning 'marsh', but observation of the extremely distinctive river-side sites to which it is applied makes it clear that it had a highly specialised meaning which is certainly not 'marsh'. The number of words needed to give an adequate account of this term in modern English illustrates the poverty of our present topographical vocabulary. The picture of Buildwas in Shropshire (Fig. 3.19), where the River Severn floods and drains spectacularly in the space of one or two days, well illustrates the nature of this element, which I have discussed more fully in *Place-Names in the Landscape* (Gelling 1984, 59–60). Since this discussion was published it has become usual to render *wæsse* 'alluvial land' or 'riverside land liable to flood' rather than 'marsh' as in earlier reference books.

Terms which really do refer to marshes include *mōr* and *fen*, which were probably not synonyms. There is no doubt that *mōr* is applied to fairly extensive areas of low-lying wet ground, as in Otmoor. The word has a dual

3: Place-names and Landscape

personality, being also used, as it is in modern speech, for barren uplands. Here again there is no agreement among philologists about the origin and ultimate etymology of the word. The other word, *fen*, is rarely used in ancient names in the parts of eastern England now called The Fens. Limited investigation of its use elsewhere suggests that it may have been a specialised term for a linear marsh, as found at Fulvens in Shere, between Guildford and Dorking in Surrey.

3.19: Buildwas SHR, with the River Severn in flood

The wetlands of Somerset and the eastern Fens were sometimes actually under water in earlier times, and it is instructive to study the locations of settlements with names containing *hȳth* in such areas. This is the place-name term for an inland port. Some *hȳth*s are on major rivers: a series runs up the Thames as far as Bolney, west of Maidenhead. Maidenhead ('port of the maidens') was the highest point to which the Thames was regularly navigable in the Middle Ages. Others, however, are sited at the junction of fen and firm ground. Bleadney is in the gap where the River Axe flows through the ridge along which the settlements of this part of Somerset are clustered, and it seems reasonable to assume that this *hȳth* had a major role

3: Place-names and Landscape

in the distribution of goods at times when land transport was not possible to north and south of the ridge (see Fig. 3.20). Aldreth is on the western tip of a large block of raised land on which settlements cluster in this part of Cambridgeshire. Aldreth is well-known to historians because of the causeway over the fen, perhaps made by William the Conqueror during the revolt led by Hereward the Wake. The name, however, speaks of the function of Aldreth in pre-Norman times. Four miles across the fen to the west is Earith in Huntingdonshire, also on the edge of firm ground. These are two examples of a large cluster of *hȳth* names which must represent a system of water transport in the early Anglo-Saxon period (see Fig. 3.21).

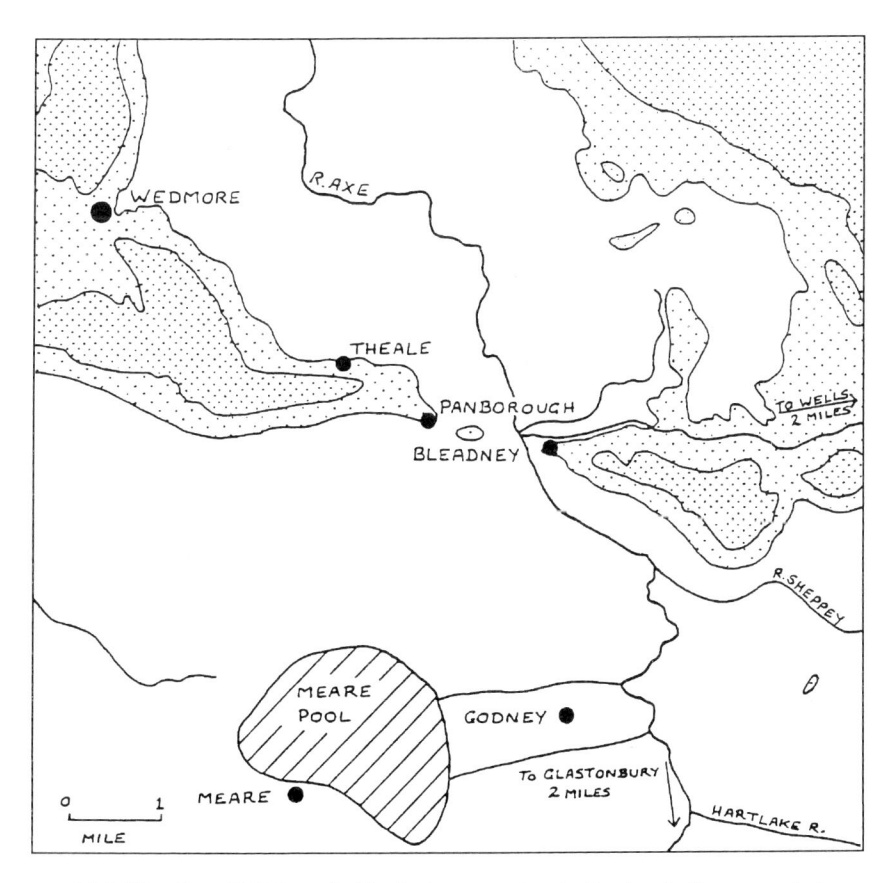

3.20: Bleadney SOM: probable drainage system in the Anglo-Saxon period
(Shaded areas over 50'.)

3: Place-names and Landscape

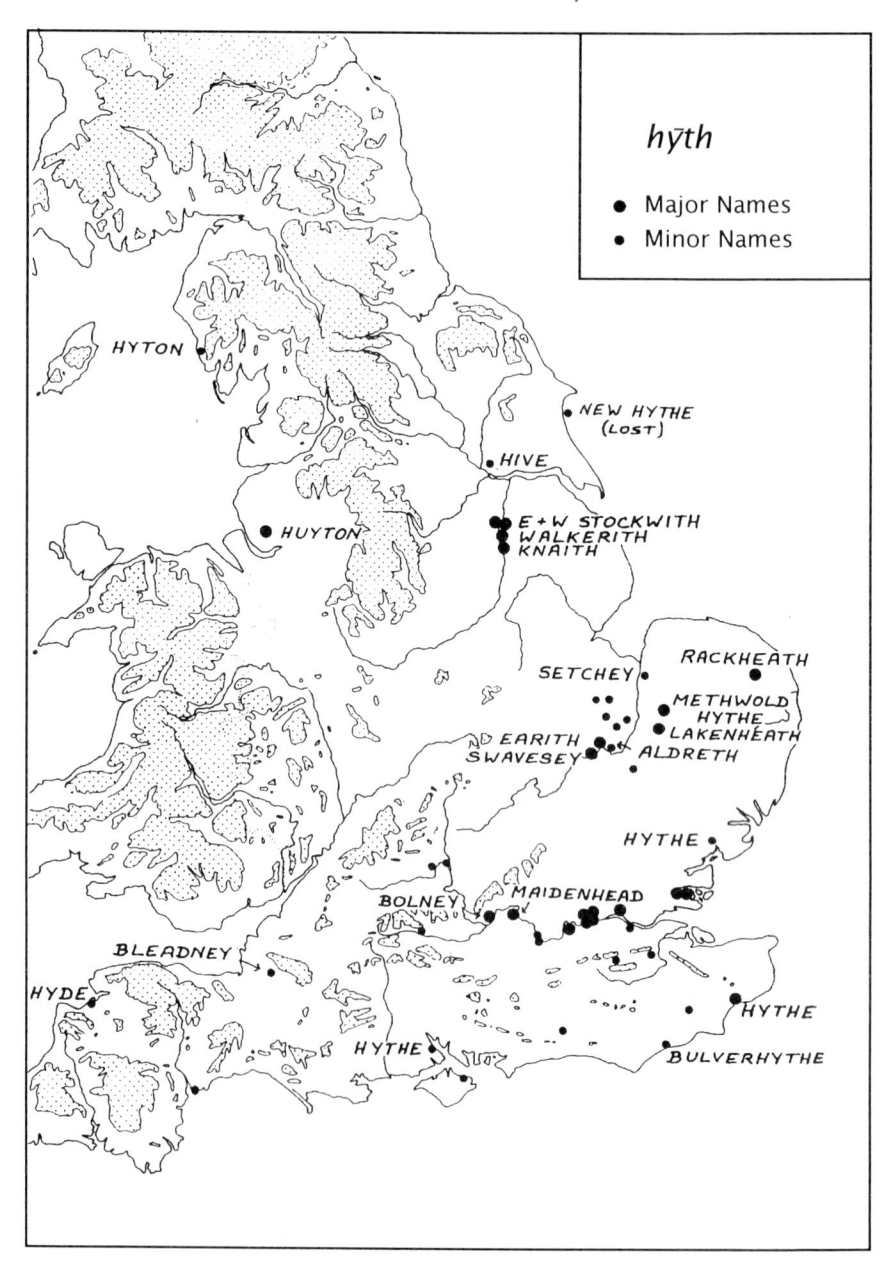

3.21: Place-names containing Anglo-Saxon *hȳth* in England

3: Place-names and Landscape

Roads also have a varied vocabulary. The word *pæth* seems to me to have been a specialised term for roads over heath and high moorland. Morpeth in Northumberland has a name meaning 'murder path', which seems inappropriate for a large market town situated where a network of roads crosses the River Wansbeck. Northumberland historians tell me, however, that Morpeth only attained this status after the Norman Conquest, when the new castle caused a shift of emphasis from Mitford, its neighbour to the west, so perhaps one of the roads to Morpeth was as lonely in pre-Conquest times as the road to Hudspeth is now, as illustrated in Figure 3.22.

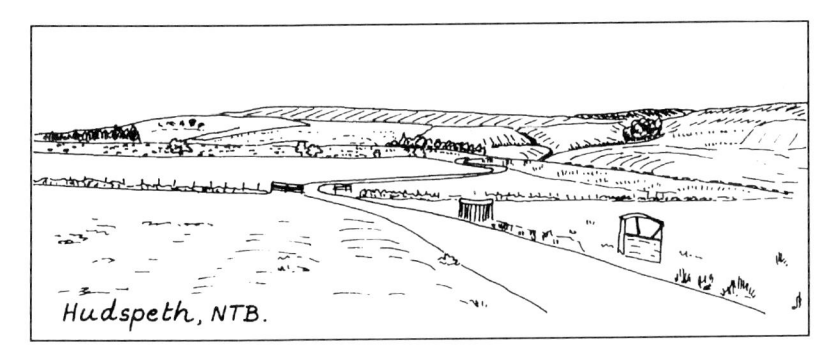

3.22: Illustration of a place with a name containing Anglo-Saxon *pæth*

Other categories of landscape features for which the Anglo-Saxons had highly specialised terms are streams, ponds, woods, and different types of meadow and pasture land. I hope that my efforts, complemented by those of Ann Cole, have rescued topographical settlement-names from the neglect of previous years and established them as an important source of information about the colonisation and exploitation of the post-Roman landscape. They afford valuable insight into the minds of Anglo-Saxon farmers, for whom a successful evaluation of their physical environment was essential to survival, and whose exploitation of the landscape produced the surplus wealth on which was based the rich culture of Anglo-Saxon England.

Bibliography
Cole, A., 'The Meaning of the OE Place-Name Element *ōra*', *English Place-Name Society Journal*, 21 (1989), 15–21.
Cole, A., 'The Origin, Distribution and Use of the Place-Name Element *ōra* and its Relationship to the Element *ofer*', *English Place-Name Society Journal*, 22 (1990), 27–41.

3: Place-names and Landscape

Gelling, M., *Place-Names in the Landscape* (1984).

Appendix

Topographic elements in English place-names discussed in the text, in order of appearance.

dūn 'hill, mountain'. Characteristically used in place-names for a low hill with a fairly extensive and fairly level summit which provides a good settlement-site. Ultimate etymology not determined. Modern English *down*. Figs. 3.1–3a–d.

>Garsington OXF, *Gersendona* c.1110, 'grassy hill'. Fig. 3.3a.
>Billington BDF, *Billendon* 1196, 'Billa's hill'. Fig. 3.3b.
>Cuddesdon OXF, *Cuthenes dune* 956, 'Cūthen's hill'. Fig. 3.3c.
>Stottesdon SHR, *Stodesdone* 1086, *Stoteresdon* 1341, 'stud hill' or 'stud-keeper's hill'. Fig. 3.3d.

beorg 'mountain, hill'. Regularly used in place-names for small, continuously curving hills, sometimes occupied by single farms or by village churches, frequently having the village beside them. Also used for burial mounds in the southern half of England. In areas of Norse settlement it is often impossible to say whether OE *beorg* or ON *berg* is involved. Cognates in Indo-European languages with basic meaning 'height'. Modern English *barrow*. Fig. 3.4a–d.

>Gran(d)borough BUC, *Greneberge* c.1060, 'green hill'. Fig. 3.4a.
>Bromsberrow GLO, *Bremesberge* 1221, 'Brēme's hill'. Fig. 3.4b.
>Edlesborough BUC, *Edulfesberga* 1175, 'Ēadwulf's hill'. Fig. 3.4c.
>Rook Barugh YON, *Rocheberch* 1160, 'rook hill'.
>Birkland Barrow LNC, *Berchlundberghe* c.1225, 'birch-copse hill'.
>Scaleber LNC, *Scaleberge* 1202, 'shieling hill'.
>Brownber WML, *Brownebar* 1591, 'brown hill'.
>Roseberry (Topping) YON, *Othenesberg* 1119, 'Othin's hill'. Fig. 3.4d.

*crūg, a Primitive Welsh word, Modern Welsh *crug* 'mound'. This was adopted by Old English speakers in the forms *crȳc*, *crīc* and *crūc*. Field-work suggests that in place-names likely to be of Old English origin it denotes an isolated hill which makes a striking visual impact. Fig. 3.5a–d.

>Crookbarrow WOR, *Crokeberewe* c.1225, 'abrupt hill with a curved profile'. Fig. 3.5a.
>Creechbarrow SOM, *Crycbeorh* 682, a doublet of Crookbarrow. Fig. 3.5b.
>Crutch WOR, *Cruch* 1178, 'abrupt hill'. Fig. 3.5c.
>Crouch Hill WLT, 'abrupt hill'. Fig. 3.5d.

hōh 'heel', used in some regions, such as East Anglia, of any spur of land, but applied where the topography permits to spurs of a characteristic shape. These rise to a point and fall away abruptly with a slightly concave curve. This resembles the foot-shape of a person lying face-down. Fig. 3.6a–c.

>Ivinghoe BUC, *Evingehou* 1086, 'hill-spur of Ifa's people'. Fig. 3.6a.

3: Place-names and Landscape

Tysoe WAR, *Tiheshoche* 1086, 'Tiw's hill-spur'. Fig. 3.6b.
Weo SHR, *Weho* 1225, 'hill-spur with road'. Fig. 3.6c.
Rainow CHE, *Rauenouh* 1285, 'raven's hill-spur'.
Ingoe NTB, *Inghou* 1242, first element uncertain. Fig. 3.6d.

ofer, only evidenced in place-names, but long recognised as an established lexical item and given the general meanings 'slope, hill, ridge'. Field-work establishes that it refers to level ridges with convex shoulders. Figs. 3.7a–d, 3.10–3.11.

Haselor WAR, *Haseloura* 1150, 'hazel ridge'. Fig. 3.7a.
Shotover OXF, *Schotouer* 1142, 'ridge with a steep ascent'. Fig. 3.7b.
Wellingore LIN, *Wellingoure* 1086, 'spring-dwellers' ridge'. Fig. 3.7c.
Wentnor SHR, *Wantenoure* 1086, 'Wenta's ridge'. Fig. 3.7d.

ōra. This word, a borrowing from Latin, is used in the south of England to denote a feature comparable to that called **ofer*. Figs. 3.8–3.11.

Chinnor OXF, *Chenore* 1086, 'Ceonna's ridge'. Fig. 3.8.
Pershore WOR, *Perscoran* 972, 'osier ridge'. Fig. 3.9.

denu 'valley', used in place-names for long valleys, usually curving, with a gentle gradient which offers a good course for a road. The *Oxford English Dictionary* says 'root-meaning uncertain'. Old Norse equivalent is *dalr*, and the two words frequently interchange in place-name spellings. Areas subject to Norse influence have, on the whole, more rugged topography, so *dales* are likely to be more dramatic than *deans*. Figs. 3.12–3.13a–b.

Assendon OXF, *Assundene* c.900, 'ass valley'. Fig. 3.13a.
Standen WLT, *Standene* 778, 'stone valley'. Fig. 3.13b.
Oxendean SSX, *Oxendeane* c.1580, 'ox valley'. Fig. 3.13c.

cumb 'valley', used in place-names for short valleys shut in on three sides, characteristic of chalk and limestone country. Perhaps Old English *cumb* 'vessel', with influence from Welsh *cwm* 'deep, narrow valley; hollow'. Fig. 3.14a b.

Encombe DOR, *Hennecumbe* 1244, 'bird valley'. Fig. 3.14a.
Whitcombe (in Corton Denham) SOM. Fig. 3.14b.
Castle Combe WLT, *Cumbe* 1188.

hop, apart from place-names only evidenced in *Beowulf*, where it means 'lair'. Ultimate origin not known. Much the commonest place-name usage is for very secluded valleys, and concentrations occur in north Herefordshire/south Shropshire, the edges of the Pennines and the Scottish borderlands. Figs. 3.15–3.17.

Hope Dale SHR. Figs. 3.15–3.16.
Stanhope PEB, *Stanehoip* 1564, 'stone valley'. Fig. 3.17.
Cowpe LNC, *Cuhope* c.1200, 'cow valley'.
Oxnop YON, *Oxenhop* 1301, 'oxen valley'.

ēg 'island', characteristically used in place-names to denote slightly raised areas which

3: Place-names and Landscape

provide dry settlement-sites in marshy or easily-flooded ground. The Old Norse equivalent is *holmr*. Fig. 3.18a–b.

    Kingsey BUC, *Eya* 1174.
    Athelney SOM, *Æthelingaeigge* 878, 'princes' island'. Fig. 3.18a.
    Middlezoy SOM, *Soweie* 725, 'island by River Sow'.
    Othery SOM, *Othri* 1225, 'other island'. Fig. 3.18b.

wæsse, only known from place-names; a highly specialised term for alluvial land which floods and drains very quickly. Fig. 3.19.

    Buildwas SHR, *Beldewes* 1086, first element possibly 'surging'.

mōr and *fen*, two 'marsh' terms.

    Otmoor OXF, *Ottemore* 1340, 'Otta's marsh'.
    Fulvens SUR, *Fulefenne* 1241, 'foul marsh'.

hȳth 'inland port'. Figs. 3.20–3.21.

    Bleadney SOM, *Bledenithe* 712, 'Bledda's port'.
    Aldreth CAM, *Alrehetha* 1170, 'alder port'.
    Earith HNT, *Herheth* 1244, *Earheth* 1260, 'gravel port'.

pæth 'path', frequently used in place-names for roads over heath and moorland. Fig. 3.22.

    Morpeth NTB, *Morthpath c.*1200, 'murder path'.
    Hudspeth NTB, *Hodespeth* 1252, 'Hod's path'. Fig. 3.22.

4

Place-name Distributions and Field Archaeology in South-west Wales

Terrence James, FSA

Archaeologists have always considered place-name studies part of their stock-in-trade. However, since archaeology is such a multi-disciplinary subject, archaeologists have become preoccupied with new methodologies and the benefits that technology has brought. One has only to think of the revolution brought about by dating methods like radio-carbon[14], tree-ring analysis, or new processes like remote sensing, geophysical survey and air photography to realise how science has broadened the tools available. To some extent archaeologists have been content to leave place-names to place-name scholars.

This chapter will therefore cover some fairly straightforward ground, looking at the type of place-name elements that are quite obvious and numerous. On the face of it Welsh place-names appear fairly transparent, for unlike those, say, in England, many Welsh names are for the most part intelligible to Welsh speakers. However there are still many Welsh place-names whose meanings are obscure or difficult of interpretation. The elements that I examine here are a selection which indicate the existence, or former existence, of archaeological sites.

The structure of archaeological organisation in Wales differs from both Scotland and England. In Wales the body charged with the care of monuments and listed buildings is Cadw (the equivalent of Historic Scotland). The Royal Commission on the Ancient and Historical Monuments of Wales (like its sister body in Scotland) has prime responsibility for recording and record keeping. At a regional level in Wales there are four Archaeological Trusts which were set up in 1974–5 with a prime task of undertaking rescue archaeology. An early priority for some Trusts was the establishment of Sites and Monuments Records (SMRs). These Records were initially based on information held on record cards compiled by the now defunct Archaeology Section of the Ordnance Survey, which included some place-name evidence. I had a close involvement with the establishment of the SMR in Dyfed (the south-west of Wales formed from the historical and recently re-created counties of Carmarthenshire, Cardiganshire and Pembrokeshire). The SMR was designed for computerisation at the outset, and records were created by scanning local and national journals, museums accessions' registers and any publications that may have contained information about sites or finds. Subsequently manuscript maps and tithe schedules were searched for place-names suggestive of archaeological sites. This chapter will rely heavily on information contained within the Dyfed SMR which currently contains about 30,000 sites and will be supplemented by data from the Carmarthenshire Place-name Survey, which contains about 35,000 forms.

South-west Wales is linguistically split as a result of the Anglo–Norman settlement pattern. In south-west Pembrokeshire – often called 'Little England beyond Wales' – Welsh was virtually exterminated by the thorough Flemish and Anglo–Norman settlements during the reign of Henry I (1100–35), and many of the distribution maps clearly show this. Elsewhere Welsh place-names predominate in areas where Welsh is still widely spoken.

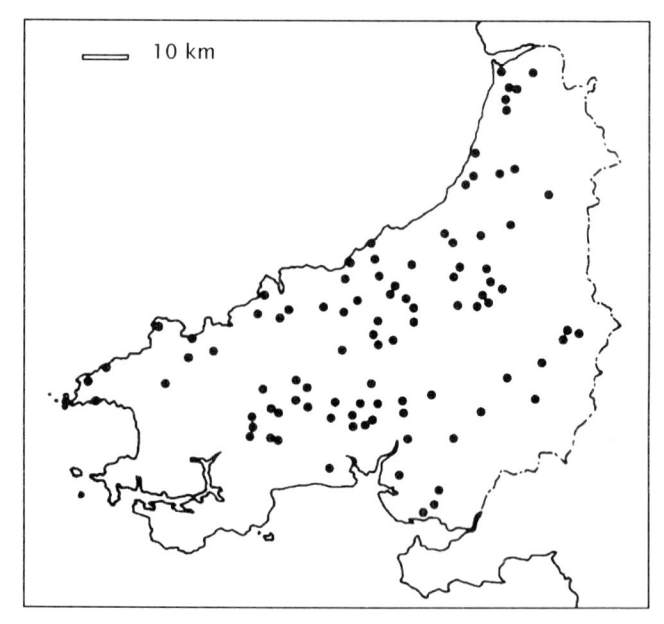

4:1: *caer/gaer*-named hillforts in Dyfed, South-west Wales
(Map: J. Renny)

4: Place-name Distributions and Field Archaeology in SW Wales

I will start by considering place-names indicative of defended settlements of the late prehistoric period (the Iron Age) through the Roman period to the Middle Ages. There are currently 732 hillforts recorded in Dyfed, of which 393 have names indicative of defence. The remaining 389 have either no known name, or names relating to local topography or of recent origin. The first element to consider is *caer* (Fig. 4.1). This word, from the same root as *cau* 'to enclose', is a very common element found in a number of town names and is to some extent analogous to the *chester* name in England, which indicates former Roman forts and towns.

The pronunciation is sometimes altered to 'car' as in Cardiff or Cardigan when the words are anglicised. The *caer* element is recorded in 100 of the 732 hillforts but surprisingly only one of the four Roman fort sites in Dyfed is so called – Carmarthen. The remaining forts on this distribution map are Iron Age or just possibly Early Medieval in date. The absence of *caer* names in south Pembrokeshire is absolute. A pitfall for the collector of the *caer* element is that *cae* is the word for field so that 'the field of the oak' might translate as *Cae y dderwen* compressed to *Cae'r dderwen*: if the apostrophe is dropped then we are in trouble! Usually such contractions are fairly obvious, but the non-Welsh speaker needs to be cautious. *Caer* is rarely applied to later castles: Figure 4.2 shows Carew, a fine thirteenth-century castle converted into a Tudor residence. Air photographs taken from 1949 show that beneath the medieval castle lies a multi-vallate Iron Age promontory fort. It has been argued that the name Carew is not composed of the *caer* element. But the confirmation of the existence of the promontory fort in excavation now allows us to confidently say that Carew is derived from *caerau* the plural of *caer*. Multi-vallate forts – those with many ramparts and ditches – are often called *caerau*. I will refer to another example later.

The next element to consider is *castell* and *castle*, both from the same Latin root (Fig. 4.3). Generally speaking the word is thought to indicate a medieval fortification, although since there are 129 hillforts with a *castell* element and 52 with *castle* it is clear that prehistoric fortifications are often called *castell/castle*. An absence of *castell* sites on the distribution map can be noted in south Pembrokeshire.

The third element to be considered is *din* or *dinas* (Fig. 4.4), which is comparable with the Gaelic *dùn* as is Dundee or Dunfermline meaning 'fort' or 'hill top'. The distribution of *din* is much thinner than the others with only 30 hillforts containing this element. The sites are predominantly in the north and sparsely distributed along the southern seaboard. There are also a number of Welsh towns which contain the element: Tenby and Denbigh are from *dinas fach* 'little fort'. But Carmarthen's *din* element has to be dug out

4.2: CAREW CASTLE, Carew parish PEM (SN044037), with crop marks showing evidence of the former prehistoric promontory fort in the foreground
(left of the walled garden)
(Crown Copyright: Royal Commission on the Ancient and Historical Monuments of Wales)

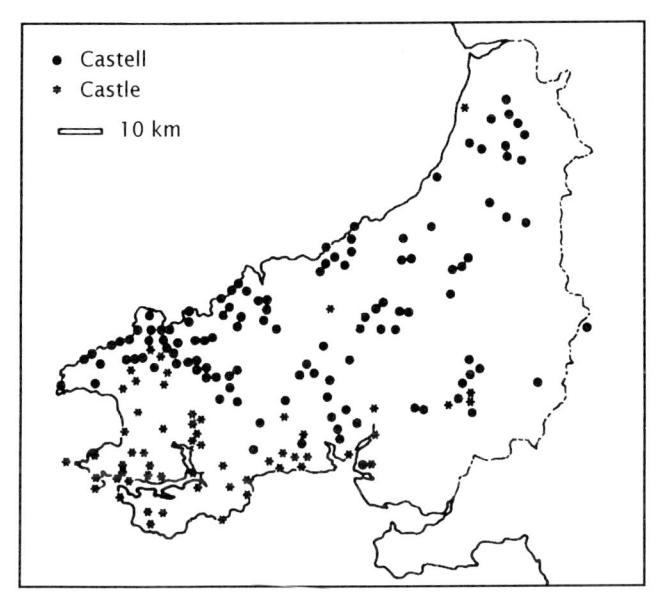

4: Place-name Distributions and Field Archaeology in SW Wales

4.3: Distribution of *castell* and *castle* place-names in Dyfed
(Map: J. Renny)

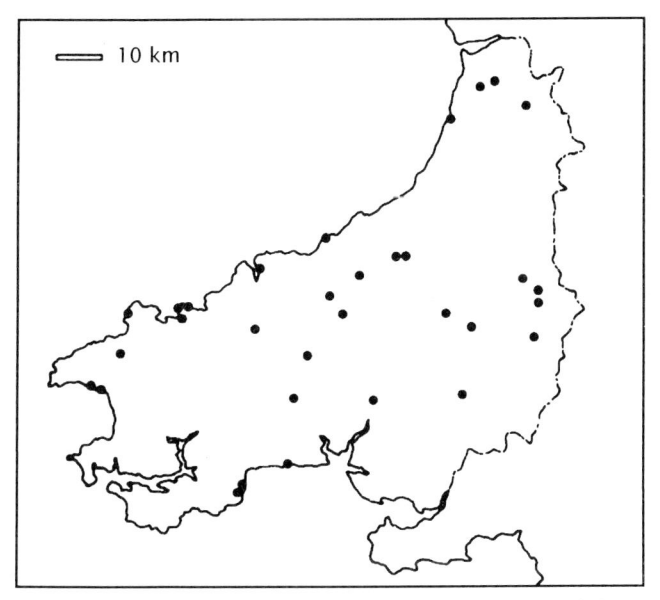

4.4: Distribution of *din* and *dinas* place-names in Dyfed
(Map: J. Renny)

from its original Roman name *Moridunum*: if we remove the later prefix *caer* from Carmarthen we are left with two elements: Brythonic **mori-* ('sea') and *dunon* ('fort or city'), thus it means 'sea fort'. The *caer* prefix is a later post-Roman addition (Jarman 1986, 24). In the context of hillforts however, the *din* element is almost universally placed first or qualified by an adjective (for a recent discussion of *din* see Moore 1993).

Of the distributions so far we have seen virtually no sites in south Pembrokeshire, but I can assure you that there are hill and promontory forts there (Fig. 4.5). We can largely adjust the anomaly by looking at *rath* as an element (uncertainly related to the Welsh *rhath*), and comparable to the Gaelic *ràth* 'fort' (Charles 1982, 40; Charles 1992, ii, 809; Williams 1945, 60). There are about 40 *rath* sites, all but one confined to the lordships of Pembroke, Rhos and Daugleddau (the last two being the principal areas of Flemish settlement). However the low numbers in the Lordship of Pembroke (the southernmost peninsula) clearly calls for further investigation. Had the *rath* element been an inheritance from the post-Roman Irish invasions of west Wales, then one would have expected to see many examples in north Pembrokeshire and the adjacent counties. The name must therefore be a later import. The distribution coincides so closely with the area of Flemish settlement that it is tempting to suggest that the element was introduced by people like Wizo Flandrensis, leader of the Flemings, who gave his name to Wiston PEM (Toorians 1990). However there are few *early* examples of the *rath* element in B.G. Charles' *Place-names of Pembrokeshire*. The earliest he cites are dated 1603 and 1651; the majority of forms are first recorded in the tithe apportionment so we need to have more evidence yet to dismiss the possibility that the element is a late, perhaps antiquarian, coinage encouraged by 'little England beyond Wales' enthusiasts! If we compare *rath* with the English element *-ton* or *-tun* denoting a farmstead (Fig. 4.6) we can see how *-ton* has a more northern boundary and a much more numerous distribution in the south, underlining the confined distribution of *rath*.

There is one other name element denoting a fortification: *berry*. There are only three, and these in areas of English settlement. The word *cadair*, literally 'chair', which is often used for high points, is not used for hillforts in Dyfed, unlike in north Wales.

Scholars accept that the chronology of what may be termed 'defensive elements' is that *caer* replaces *dinas*, then *castell* replaces *caer*. Looking at the overall distributions, what can be said about the various elements already mentioned? The main point to make is that hillforts with elements suggestive of defence – *caer*, *dinas* and *castell* – are in areas that were not, or were only moderately, settled by the Anglo–Normans (the Welshries). This can be best

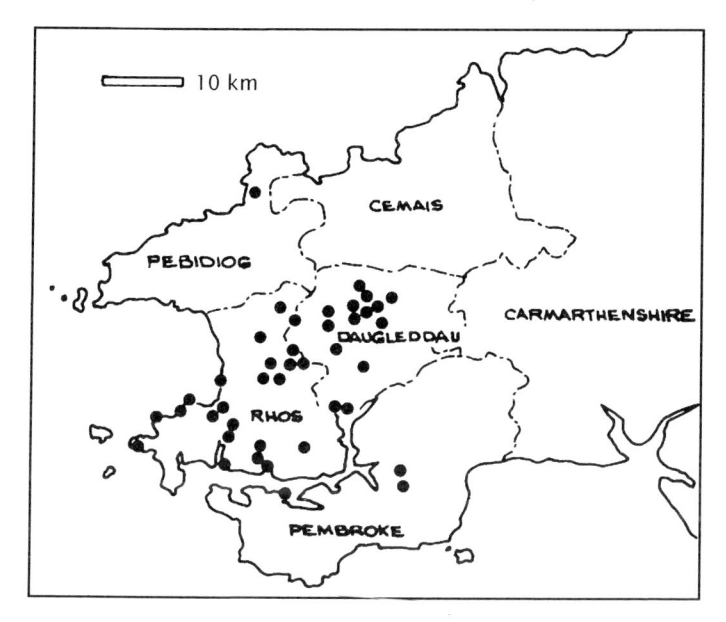

4: Place-name Distributions and Field Archaeology in SW Wales

4.5: *rath* in Pembrokeshire showing medieval Lordships (Map: J. Renny)

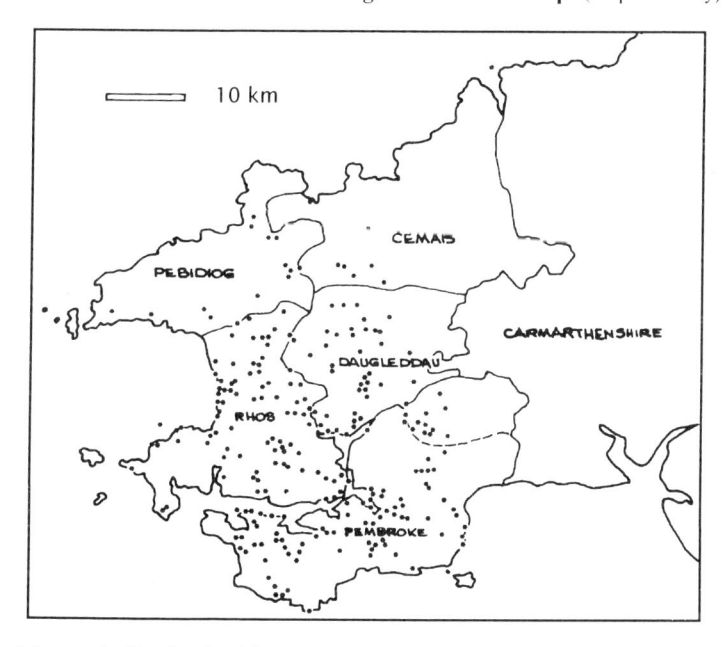

4.6: *-ton* in Pembrokeshire showing medieval Lordships (Map: J. Renny)

4: Place-name Distributions and Field Archaeology in SW Wales

seen by looking at the distribution of defensive sites *without* defensive names (Fig. 4.7): here there is a concentration towards the areas of Anglo–Norman settlement in the south-west. The distribution of the *castle* element has the same geographical bias (Fig. 4.3): indeed hillforts with *castle* names are almost exclusively confined to the richer lordships of Pembrokeshire and south Carmarthenshire (the Englishries). It would appear that linguistic dislocation has caused the original (or at least earlier) names for most forts to have been lost in these areas. The distribution of hillforts with *castle* names reinforces the point (Fig. 4.8).

Where there is no known archaeological site a place-name like *caer* or *castell* has often lead to new discoveries when followed up by field work or aerial photography. Figure 4.9 shows Castell Draenog ('Hedgehog Castle'): until the dry summer of 1984 no site was known here other than the place-name, but the cropmarks are clear evidence of a lost bi-vallate hillfort.

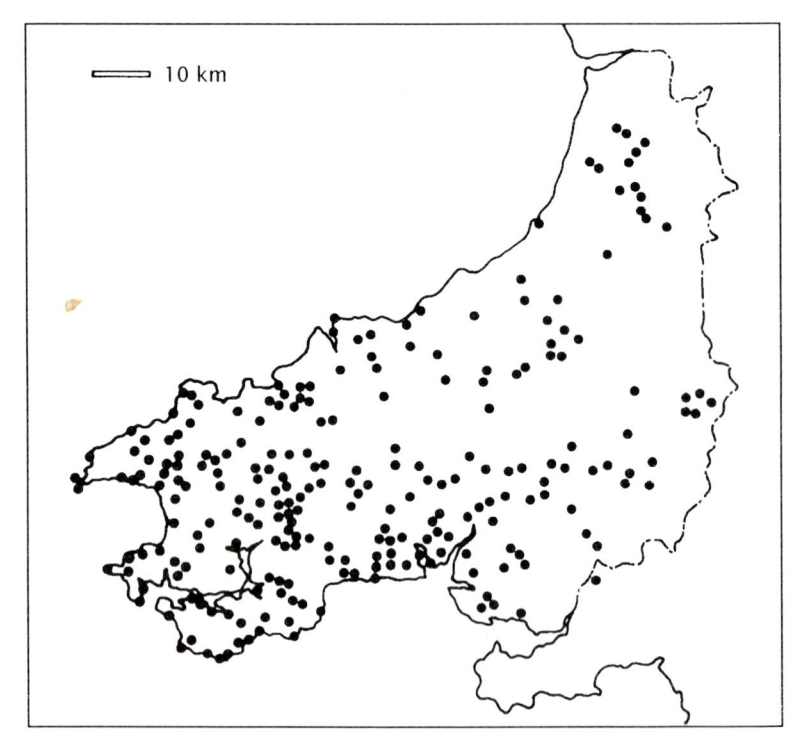

4.7: Hillforts in Dyfed not named *caer/castell/castle/rath*
(Map: J. Renny)

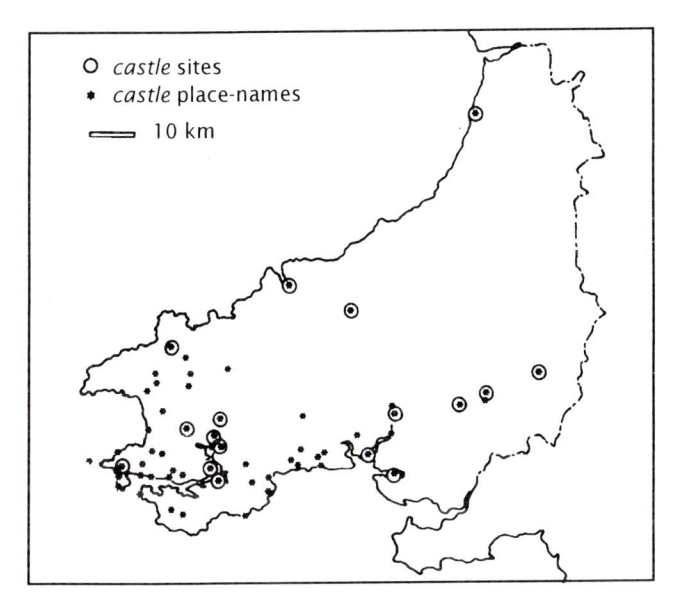

4: Place-name Distributions and Field Archaeology in SW Wales

4.8: Distribution of *castle* place-names in Dyfed (Map: J. Renny)

I have looked at *castle/castell* sites which contain a personal name. The distribution of these cannot be used to demonstrate anything meaningful. I have also investigated the possibility that defended enclosures with a *castle/castell* element could indicate a post-Roman use, if not a post-Roman origin. This too has proved inconclusive. In west Wales there is a class of small curvilinear enclosure that is contemporary with motte and bailey castles. They are known as ringworks. Some of these were subsequently converted into stone castles: a good example is the castle of the bishops of St Davids at Llawhaden. However, some ringworks are undoubtedly prehistoric, but it is not always obvious from what period these sites originate. Another means of testing the hypothesis that *castell/castle* names might give a clue to date, is to see if they are found in close proximity to churches or chapels. Such a pairing could form the joint focal point of a planted medieval settlement. In this case there are sufficient instances of paired churches and *castell*-named ringworks to support the contention (Fig. 4.10). Each pair of overlying symbols in the distribution map represents a church or chapel within 500m of a *castell/castle*-named hillfort. We can reasonably deduce from this that many *castell/castle*-named ringworks are medieval or are earlier fortifications reused to form the *caput* of planned settlements.

4.9: The site of CASTELL DRAENOG, Llanboidy parish CRM (SN213214), showing as a bi-vallate cropmark in grassland (Copyright Dyfed Archaeological Trust)

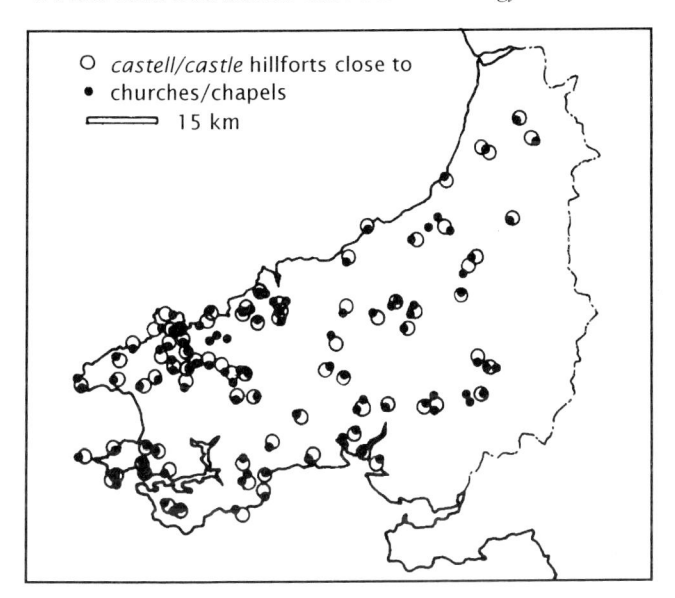

4.10: *castell/castle* hillforts in Dyfed close to churches and chapels (Map: J. Renny)

I now turn to place-names associated with ecclesiastical sites and settlement. The most common element in Welsh parish names is the word *llan* usually prefixed to a founder-saint's name: Llangynog CRM, which has an archetypal rounded *llan* churchyard, is the church of St Cynog. Of the 1,000 plus parishes in Wales the vast majority start with *llan*. Whilst *llan* is nowadays interpreted as meaning 'church', in origin it referred to the enclosure or graveyard which may or may not have included a church. The Welsh word for church is *eglwys*, which derives from Latin *ecclesia* and is therefore closely related to the *eccles* place-names of England and Scotland. It has been argued by Kenneth Cameron for England (Cameron 1968, 90–1) and Geoffrey Barrow for Scotland (Barrow 1983) that such *eccles-* names indicate a very early church foundation. Tomos Roberts has more recently argued the same for *eglwys* names in Wales (Roberts 1992). In the area of south-west Wales under discussion there are a few surviving *eglwys* churches: Eglwys Gymyn, Eglwys Wrw, Eglwys Fair a Churig and Eglwys Cyffig. Some of these churches are undoubtedly on ancient sites. Eglwys Gymyn has a classic round *llan*-type churchyard, which is considered to have developed from a prehistoric enclosure. It has a sixth-century class 1 Early Christian Monument with both Latin and Ogam inscriptions. Tomos Roberts also put forward the suggestion that there may only have been one *eglwys* church in a commote (*ibid*). The distribution of *eglwys* sites (Fig. 4.11) has therefore been placed over the commotal boundaries: his assertion is generally true, although as we can see there are two extant churches in the commotes of Laugharne and Pembroke. Compounding the problem is the fact that *eglwys* names tend to be replaced by *llan*. We can see this in the following example of Newchurch, today called Llan-newydd but originally Eglwys Newydd. Newchurch is quite well documented because it was granted to Carmarthen priory in the twelfth century (Phillips 1865):

Date	Name
1147–1176	*Eglusnewith*
1154–1186	*Eglusnewit*
1247	*Eglisuent*
c.1340	*Egloysnewith*
1395	Capella de *Nuecherche*
1786	*Llan Newydd* alias *Newchurch*
1913	*Llannewydd*

Another parish church close to Carmarthen – Llangain was *Egliskein* in 1247 but had become *Langayng* by 1552.

[Sources: Phillips 1865, Docs. 32–3, 119; Daniel-Tyssen 1878, pp. 4–5, 35; Baker 1907; Arch. Camb, 1913, 399.]

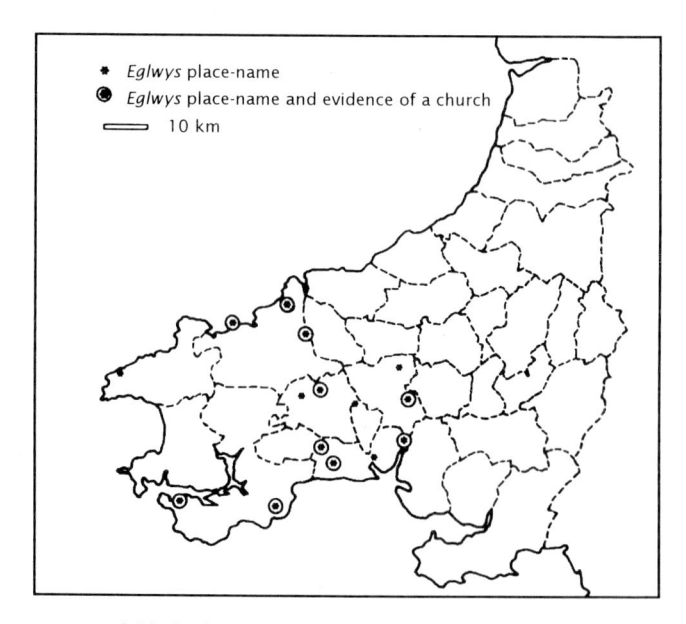

4.11: Dyfed, showing commotes (Map: J. Renny)

Eglwys names also survive where no church is to be found, but where archaeology combines with place-name evidence to indicate that a church once existed (see Fig. 4.11). A good example is the hillfort known as Caerau, near St Dogmaels PEM. (SN124454) a multi-vallate fort of Iron Age origin. One of the fields is called *Yr Eglwys Ddiflodau* – 'the church without flowers' (James, T. 1992, 65). Another is called *mynwent*, a particularly helpful element which indicates a graveyard. Between the ramparts long cist graves have been discovered providing tangible evidence of burials from the graveyard of this long-lost church. The combined evidence points to the hillfort being reused as a graveyard with a church. In another case, at Y Gaer, Bayvil PEM we have an excavated *caer* site with a cemetery which never had a church (James, H. 1992, 94–6).

Two other elements indicative of church or cemetery sites are *bedd* (grave) and the above-mentioned *mynwent*. The distribution of *mynwent* and *bedd* often occurs in conjunction with other evidence, like the record of skeletons or cist graves being unearthed, suggesting a cemetery as in the Caerau example already discussed.

The distributions (not illustrated) show a distinct north Pembrokeshire bias. This is perhaps because of the tradition there of burying in stone cist graves which are more noticeable than graves with wooden or no coffins. The combination of place-name evidence and air photography has been fairly productive in early church studies (James, T. 1992). Figure 4.12 shows the site is of an ostensibly univallate Iron Age enclosure. However in the dry summer of 1984 I noticed that there was a fainter outer enclosure of what we presume is a trench for a palisade or wattle fence. The adjacent farm is called Lan, on early maps *Llan*. One of the fields is called *Parc y Fynwent* ('Cemetery Field'), so it is possible that here we have another prehistoric enclosure being reused as a cemetery. The place-name evidence tends to support this. But does the possible *llan* name in this case point to the existence of a church or is it referring to the enclosure? It is possible that in origin *llan* referred to a graveyard enclosure without a church, while *eglwys* indicated one with a church. I have previously suggested that the thin cropmark of the palisade could be what is called a *bangor* in parts of Wales (James, T. 1992, 67). In origin *bangor* can mean 'wattled enclosure'. The *bangor* place-name is associated with early Christian sites (Jones 1993).

Air photography has been successful in discovering numerous cropmark sites with a similar morphology (James, T. 1990a), which scholars accept are prehistoric in origin, so sites with an outer palisade enclosure, arguably a *bangor*, appear to have a prehistoric pedigree.

4: Place-name Distributions and Field Archaeology in SW Wales

4.12: Lan, Llanboidy parish CRM (SN216205)

The central strong cropmark of the enclosure's ditch can be compared with faint outer enclosure for a palisade trench possibly for a wattle fence (Copyright Dyfed Archaeological Trust)

Another discovery in 1984 was the faint cropmark of a defended enclosure on a farm called Cilsant: the name can possibly be interpreted as 'saint's cell'. An adjacent field produced a possible inscribed stone (now unfortunately lost). Cilsant was the residence of Cadifor Fawr 'the supreme lord of Dyfed' who died in 1091. He was the father of Bledri Latimer the supposed purveyor of the Grail stories and benefactor of Carmarthen Priory (James, T. 1992).

Before ending I wish to go back in time to the Romans. Wales is dotted with place-names containing the element *sarn* 'raised embankment' or *'agger'*. Many known Roman roads are called *Sarn Helen*, erroneously named after Helen the mother of Emperor Constantine, who was somehow conflated with the legendary British wife of Magnus Maximus (Macsen Wledig). The distribution map of the *sarn* element from the Carmarthenshire Place-name Survey (Fig. 4.13) is partly self-fulfilling in that the nice straight line of dots running north from the central square symbol is based on *Sarn Helen* names recorded (or created?) by the Ordnance Survey along a known road line. We can see this when the line of known roads is overlaid on the distribution. The square in the middle is Carmarthen, site of a fort and the *civitas* capital of the Demetae (the Iron Age tribe that gives its name to Dyfed). In the conventional wisdom there are no Roman roads west of Carmarthen, but significantly there are several *sarn* elements. In 1990, whilst perusing a set of vertical air photos taken in 1983, I discovered a straight cropmark continuing the line of an otherwise deviating road which looked very like the sort of evidence one looks for in a Roman road. By following this line back to Carmarthen from photographs taken on an earlier sortie of 1955 it was possible to trace a linear cropmark in total for 17 kilometres (James, T. 1990b, 55–6). Subsequent field work and excavation has proved this to be a completely new road leading westward into Pembrokeshire, a county apparently devoid of forts and Roman incursions. The road line, marked as a pecked line in Figure 4.10, shows some *sarn* place-names along its route. There is also a farm called Pont Cowin (i.e. 'bridge over the River Cywyn'). No bridge exists today, but the place-name recalls the bridge that perhaps spanned the River Cywyn in Roman times. To the north are clusters of other *sarn* place-names which may point to another road running northwest. No road is known along this apparent line, but there is also a tantalising place-name – a farm called Caer-lleon near the route (arrowed in Fig. 4.13). Caer-lleon means 'fortress of the Legion'. There are two other places called Caerleon in Wales, the town of that name in Monmouthshire and, in the north-east, Chester's Welsh name is *Caerlleon*. These were the principal legionary fortresses used for the subjugation of what is now Wales.

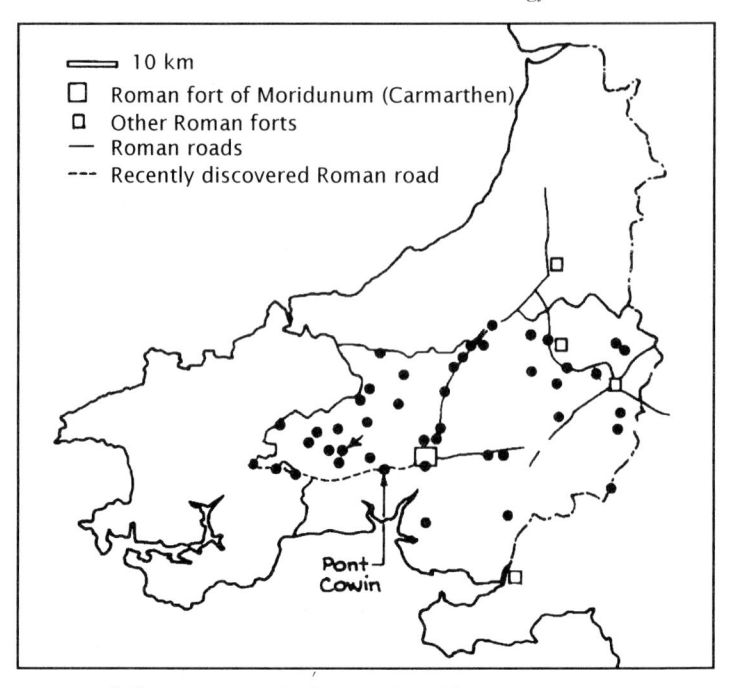

4: Place-name Distributions and Field Archaeology in SW Wales

4.13: *sarn* names in Carmarthenshire (Map: J. Renny)

The combination of place-name research, field work and air photography may uncover yet another road and perhaps a legionary fortress.

The subtleties of the distribution of the *rath* element need much more examination, and certainly more documented early forms. But what we already have points to a particular dialect word within a part of south Pembrokeshire (see p. 106 and Fig. 4.5 above). Other distributions point to other distinct dialect differences that have not yet emerged in linguistic studies. I noticed this when I mapped the distribution of the element *ynys* (Fig. 4.14). *Ynys* means 'island' but can, like the words *holm* (from Old Norse) and *innis* (in Scottish Gaelic) signify 'island in a river' or 'flat lowland liable to flood' (Williams 1945, 36–7). The element is more meaningful when plotted against rivers. There is a clear east Carmarthenshire distribution. In this context it is thought that *ynys* indicates meadow-land farmed in common, perhaps like *lammas*,[1] but why is it not present elsewhere? The answer is probably to be found in a dialect division, similar to that so well defined in one of the first place-name distributions I undertook by computer. Figure 4.15 shows the distribution of the *cae*

4: Place-name Distributions and Field Archaeology in SW Wales

element from the Sites and Monuments Record. *Cae*, as already noted, means 'field'. The east-west division is comparable to the *ynys* distribution (the boundary seems to be north-south along the rivers Tywi and Cothi – cf. Fig. 4.14). West of this line the word *parc* (from English *park*) is used. The conspicuous absence of both these elements in south-west Pembrokeshire ('Little England beyond Wales') is simply because Welsh was eradicated from this area early in the Norman conquest.

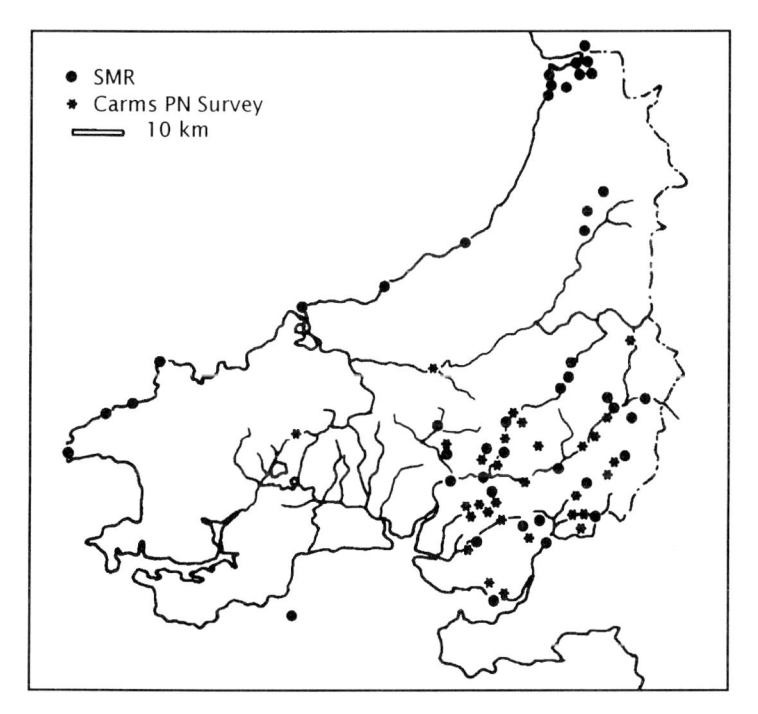

4.14: Distribution of *ynys* place-names in Dyfed (Map: J. Renny)

In this chapter I have tried to give an indication of some of the ways place-name studies can be used by archaeologists. My personal interest and profession is in the realm of computer-based applications to archaeology, and that has included the computerisation of place-names. Aware of my own inadequacies in place-name studies, I nonetheless feel that by analysing mass distributions of elements from comparatively late sources we can advance our understanding of settlement patterns, and point the way to more detailed place-name and archaeological research.

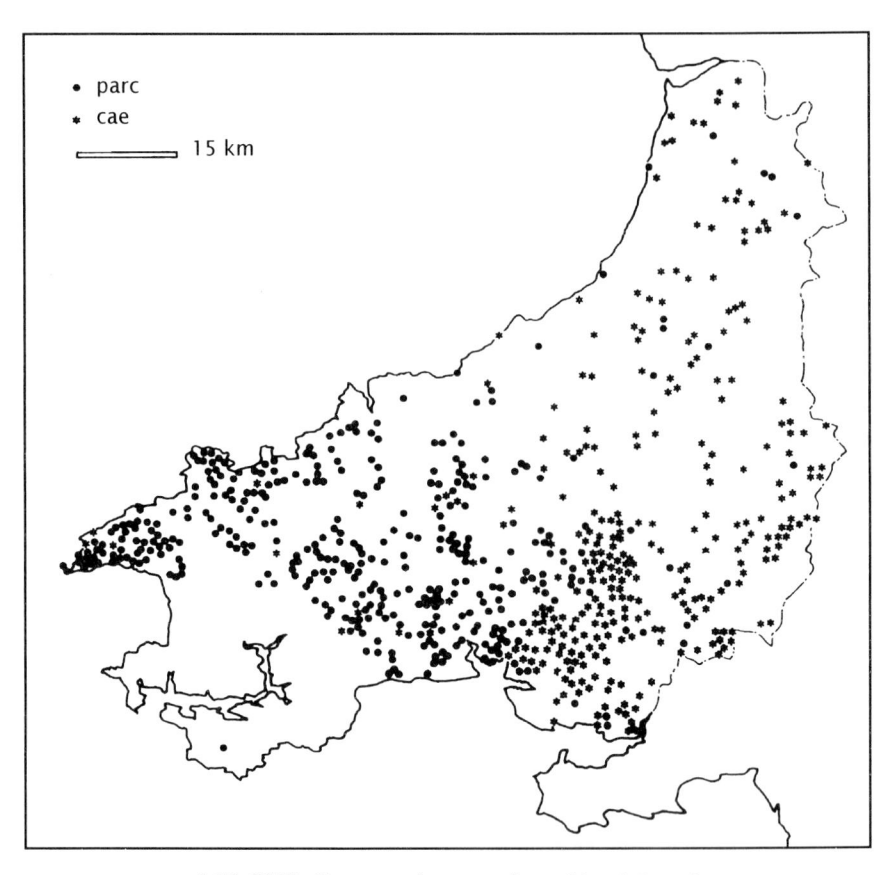

4.15: SMR sites named *parc* and *cae* (Map: J. Renny)

Note

1. Traditionally applied to meadow-land thrown open to common pasture after harvest, on or around Lammas (1 August). I am grateful to Heather James for this suggestion regarding *ynys*.

Bibliography

For a recent and constructive discussion on the linguistic boundary separating Welsh from English settlement see G.M. Awbrey 'The Term "Landsker" in Pembrokeshire', *The Pembrokeshire Historian*, 4 (1990–91), 32–44.

Arch. Camb., *Archaeologica Cambrensis*, journal of the Cambrian Archaeological Association (1847–).

Baker, T.W., *Particulars relating to Endowments etc. of Livings* (1907).
Barrow, G.W.S., 'The Childhood of Scottish Christianity: a Note on Some Place-Name Evidence', *Scottish Studies*, 27 (1983), 1–15.
Cameron, K., *Christianity in Britain, 300–700*, ed. M.W. Barley and R.P.C. Hanson (1968).
Charles, B.G., *The English Dialect of South Pembrokeshire* (Pembs. Record Society, 1982).
Charles, B.G., *The Place-names of Pembrokeshire* (2 vols., 1992).
Daniel-Tyssen, J.R., *Royal Charters... of Carmarthen* (1878).
Edwards, N. and Lane, A. (eds.), *The Early Church in Wales and the West*, Oxbow Monograph 16 (1992).
James, H., 'Early medieval cemeteries in Wales', in Edwards and Lane (1992), 90–103.
James, T., 'Concentric Antenna Enclosures – A new defended enclosure type in West Wales', *Proceedings Prehistoric Soc.*, 56 (1990a), 295–8.
James, T., 'A Roman Road West of Carmarthen?' *Archaeology in Wales*, XXX, CBA Group 2 (1990b), 55–6.
James, T., 'Air photography of ecclesiastical sites in south Wales', in Edwards and Lane (1992), 62–76.
Jarman, A.O.H., 'The Legend of Merlin and its associations with Carmarthen', *Carmarthenshire Antiq.*, XXII (1986), 15–26.
Jones, B.L., 'Why *Bangor?*', *Ainm* (Bulletin of the Ulster Place-Name Society), (1993), vol. v, 59–65.
Moore, D., 'Dinefwr and Dynevor: A Place-Name Study', *Carmarthenshire Antiq.*, XXIX (1993), 5–12.
Phillips, T., *Cartularium S. Johannis Bapt. de Caermarthen* (1865).
Roberts, T., 'Welsh ecclesiastical place-names and archaeology', in Edwards and Lane (1992), 41–4.
Toorians, L., 'Wizo Flandrensis and the Flemish Settlement in Pembrokeshire', *Cambridge Medieval Celtic Studies*, 20 (Winter 1990), 99–118.
Williams, Syr I., *Enwau Lleoedd* (1945).

Acknowledgements

I would like to thank Muriel Bowen Evans for reading a draft of this text and her helpful suggestions. I would also like to thank the St Andrews conference organisers, and Dauvit Broun for reading the paper at the conference. Simon Taylor made a number of valuable suggestions relating to sources and discussions on comparable place-name elements outside Wales, which have been incorporated into the published chapter. Lastly my thanks go to members of the Carmarthenshire Place-name Survey and the Dyfed Archaeological Trust for access to the data sources used in this article. I would also like to thank RCAHMW for permission to reproduce Figure 4.2.

5

Scandinavian Settlement in Unst, Shetland: Archaeology and Place-names

Steffen Stummann Hansen & Doreen Waugh

Part I
Steffen Stummann Hansen

Introduction

The Scandinavian phase in Shetland history includes the period from the first Viking raids around 800 AD to 1469, when the Danish King pledged the islands to the Scottish king. This phase can be divided into 'Viking' (c.800–1050 AD) and 'Late Norse' (c.1050–1500 AD). Besides the historical records, the Scandinavian era in Shetland is today extremely well represented in the place-names, of which probably approximately 95 percent are of Scandinavian origin. Also archaeology has provided evidence for the Scandinavian past, although so far to a much lesser degree than has the study of place-names. The Scandinavian settlement in Shetland has for a long time attracted interest from Scottish as well as Scandinavian specialists.

Research in Scandinavian Scotland

1996 saw the 150th anniversary of the Danish archaeologist Jens Jacob Asmussen Worsaae's pioneer journey to the British Isles and Ireland. Worsaae (1821–1885), with the support of King Christian VIII of Denmark and at the request of the Duke of Sutherland among others, conducted his visit in order to trace what might have survived of remains of the Scandinavian impact on these areas in the Viking Age and Medieval Period. A few years after his journey Worsaae published the results of his journey in the book *An Account of the Danes and Norwegians in England, Scotland and Ireland* (Worsaae 1851, 1852). The book represents a brilliant demonstration of a multidisciplinary approach to the subject, as Worsaae was interested in and recorded historical information, archaeological sites and objects and, not least, place-names. I think he deserves our tribute for that today. Although Worsaae travelled to many places in Britain and Ireland, he did not, however, visit Shetland.

Scandinavian interest in the western part of the Viking world continued, and in 1919 The Scientific Research Fund in Norway initiated an ambitious project to record and investigate the Viking remains in the British Isles and Ireland. The results were published, with the Norwegian archaeologist Haakon Shetelig (1877–1955) as editor, in six impressive volumes under the title *Viking Antiquities in Great Britain and Ireland* (Shetelig 1940–54). At the time when Shetelig and his team of Norwegian archaeologists started their impressive work, however, no settlement-sites with house foundations from the Scandinavian period were known.

Danish interest in the archaeology of the Scandinavian period of Scottish history has been limited since the days of Worsaae. However, in 1931, Aage Roussell (1901–72) from the Danish National Museum travelled in Scotland and the Western and Northern Isles in order to study remains of Scandinavian building customs in recent buildings. As a result of his journey, he published the book *Norse Building Customs in the Scottish Isles* in 1934 (Roussell 1934). A few years later Roussell was followed by the Danish human geographer Gudmund Hatt (1884–1960), who travelled to the Western Isles to search for comparative material to the houses of the Early Iron Age, which were discovered in great numbers about that time in Denmark.

Roussell describes in his introduction how the officials of the Royal Museum in Edinburgh were helpful and supportive in every respect, but at the same time had a very sceptical attitude to his project. He wrote: 'It turned out that in archaeological circles in Scotland it was the view that the Norseman always used wood as a building material, and as every relic in Scotland is of stone and earth, it cannot be of Norse origin' (Roussell 1934, 8). Further, in a work published by the Royal Commission in 1928 one can read: 'The Norsemen of the Viking period were essentially builders in wood, and no edifices of dry-built stone masonry were known in Norway either of that period or of preceding ages,' and a note adds: 'The literary evidence puts this beyond doubt' (*RCAHMS* 1928, xxiii).

The situation today is of course improved compared to the time of Roussell's visit. In northern Scotland we know of the classic site of Freswick, where V. Gordon Childe (1892–1957) and Alexander O. Curle (1866–1955) initiated and published archaeological work between the two World Wars (Childe 1943; Curle 1939, 1954). Work in Caithness has been continued within the last decades including the site of Freswick (Morris,

Batey & Rackham 1995), but still the number of located settlement sites from the Norse period is extremely limited (Batey 1987).

There are also a few partly excavated sites in the Western Isles – Drimore in South Uist (MacLaren 1974), Udal in North Uist (I. Crawford 1974, 1981, 1986; Crawford & Switzur 1977) and in the Isle of Man, for instance, Doarlish Cashen excavated in 1970 (Gelling 1971). Sheffield University has recently been working on a couple of Late Norse sites at Kilpheder (Parker Pearson & Sharples 1995; Parker Pearson et al. 1996) and Bornish in South Uist (Parker Pearson & Webster 1994; Sharples, Webster & Parker Pearson 1995; Sharples 1996).

Above all, however, a lot of research and excavation has been conducted in Orkney, especially by Durham University and latterly by the University of Glasgow (Morris 1985; Morris, Batey & Barrett 1994). The work conducted here includes not only farmsteads (Gelling 1984; Hunter, Bond & Smith 1993; Morris 1989, 1996a, 1996b; Morris et al. 1985; Ritchie 1977), but also chapel sites (Morris 1986, 1996c) and even a Norse horizontal mill (Batey 1993; Batey & Morris 1992). Mention should also be made of the Norwegian based excavations at Westness on Rousay (Kaland 1973, 1993, 1996).[1]

Scandinavian Shetland

The most famous Viking site in Scotland, however, is situated in Shetland, namely Jarlshof (Curle 1954; Hamilton 1956). This site has, since its publication, been a frame of reference for all other Norse sites to be excavated. The beginning of the Norse phases on the site was put at 800 AD by the later excavator John R.C. Hamilton (Hamilton 1956, 91–4, 106). This agreed well with the general concept of when the Viking period in the British Isles was supposed to have started. It is, however, quite clear today that there are serious problems not only with the interpretation of the excavated structures and the dating of the earliest phase at Jarlshof, but also with this early date of the Scandinavian settlement in general in northern Scotland (Bigelow 1992, 9; Morris 1985, 210–15; Stummann Hansen 1996b, 54–5). In Historic Scotland's latest guide to the site the date has been changed to 850 AD (Ashmore 1993, 12–14), but even this seems to be too early as a pin of polyhedral-headed type occurs in the earliest phase thus indicating a tenth century date (Fanning 1994, 33).

Another three Scandinavian sites have been investigated in Shetland since Jarlshof: Underhoull in Unst by Alan Small in 1963–65 (Small 1967), Da Biggins in Papa Stour by Barbara Crawford in 1977–79 (B. Crawford 1984, 1985, 1996) and Sandwick-South in Unst by Gerald Bigelow in 1978–80

(Bigelow 1985). While there can be no doubt about a Late Norse date for Da Biggins and Sandwick-South, there seems to be more uncertainty about Underhoull. The site was regarded as a 'primary Norse farmstead' by the excavator suggesting, with reference to the primary Norse phase at Jarlshof, a ninth-century date (Small 1967, 237–47). The dating of Underhoull is, however, mainly based on the lay-out of the structure and its topographical location near the sea and not on the artefact assemblage. Small already mentioned that there were allegedly tenth-century parallels to the house type in, for instance, the Faroe Islands (Small 1967, 237–47), and the characteristics of the house at Underhoull therefore cannot be confined to the ninth century. The artefact assemblage does, on the other hand, contain sherds from oval and square-sided steatite vessels indicating that it was in existence in the Late Norse period until at least the twelfth century (Stummann Hansen 1996a, 123–4).

Compared to the substantial place-name evidence, however, the archaeological evidence is not only strikingly vague in terms of number of known and excavated sites, but it also contains several chronological and interpretative problems. Anna Ritchie is therefore quite right when in her book *Viking Scotland* (1993), in a paragraph on Shetland, she states that 'understanding of the period before 1100 depends too heavily on the site at Jarlshof. Plenty of potential Viking sites are scattered across this largely undeveloped landscape, but a concentrated campaign of fieldwork and excavation is needed to confirm their Viking origin' (Ritchie 1993, 61).

It was against this background that the author, together with his colleague the Danish archaeologist Anne-Christine Larsen, initiated a pilot-project on the island of Unst in 1994. The aim of the project was to locate and describe, through a survey supplemented by a few small trial excavations, what was left of fossil remains from the Scandinavian period and thereby gain a better idea of the archaeological potential of the Shetland landscape. Unst was chosen because two large-scale excavations – those at Underhoull and Sandwick-South – had already taken place there and also because some Norse sites had been recorded previously by a native of the island, Peter Moar (1888–1983), who had been especially interested in locating the Norse settlement of the island (Fig. 5.1). In other words, we would not have to start from scratch in Unst.

Scandinavian Settlements in Unst

Unst is approximately 19 kilometres long and 9 kilometres wide. The present population of about a thousand inhabitants is concentrated in the three major settlements Uyeasound, Baltasound and Haroldswick (Fig. 5.2).

5: *Scandinavian Settlement in Unst, Shetland*

5.1: Peter Moar (1888–1983), to the right, photographed standing in the entrance of his place of work, the Danish Dairy, Commercial Street, Lerwick, in approximately 1911
(Photo: Shetland Museum)

5: Scandinavian Settlement in Unst, Shetland

The shore-line is cut by larger and smaller inlets. Around these inlets and around large parts of the shore-line low-lying areas suitable for settlement and agriculture can be seen. The inner parts of the island are mainly, and especially along the west coast, characterised by hills with a maximum height of 284 metres. The landscape of today is clearly depopulated as plenty of ruined farmsteads with adjoining field-systems and enclosures are spread all over the slopes and lower-lying areas. Many of these farmsteads were abandoned during the clearances in the eighteenth and nineteenth centuries, and even up to the middle of the twentieth century big farms have been abandoned.

5.2: View from the south-west over Haroldswick, one of the main settlements in Unst
In the foreground is the so-called King Harald's Grave. According to legend it was in this cairn that King Harald Finehair was buried after being killed during a punitive expedition to Haroldswick (hence the name). However, the king died peacefully years later in Norway
(Photo: Steffen Stummann Hansen)

The survey was conducted over a period of eight weeks during the autumn of 1994 and the spring of 1995. Remains of Scandinavian buildings were recorded on more than thirty locations (Fig. 5.3). The remains range from preserved fragments of walls to nearly completely preserved foundations of houses. In the following account some of these sites will be briefly presented

5: *Scandinavian Settlement in Unst, Shetland*

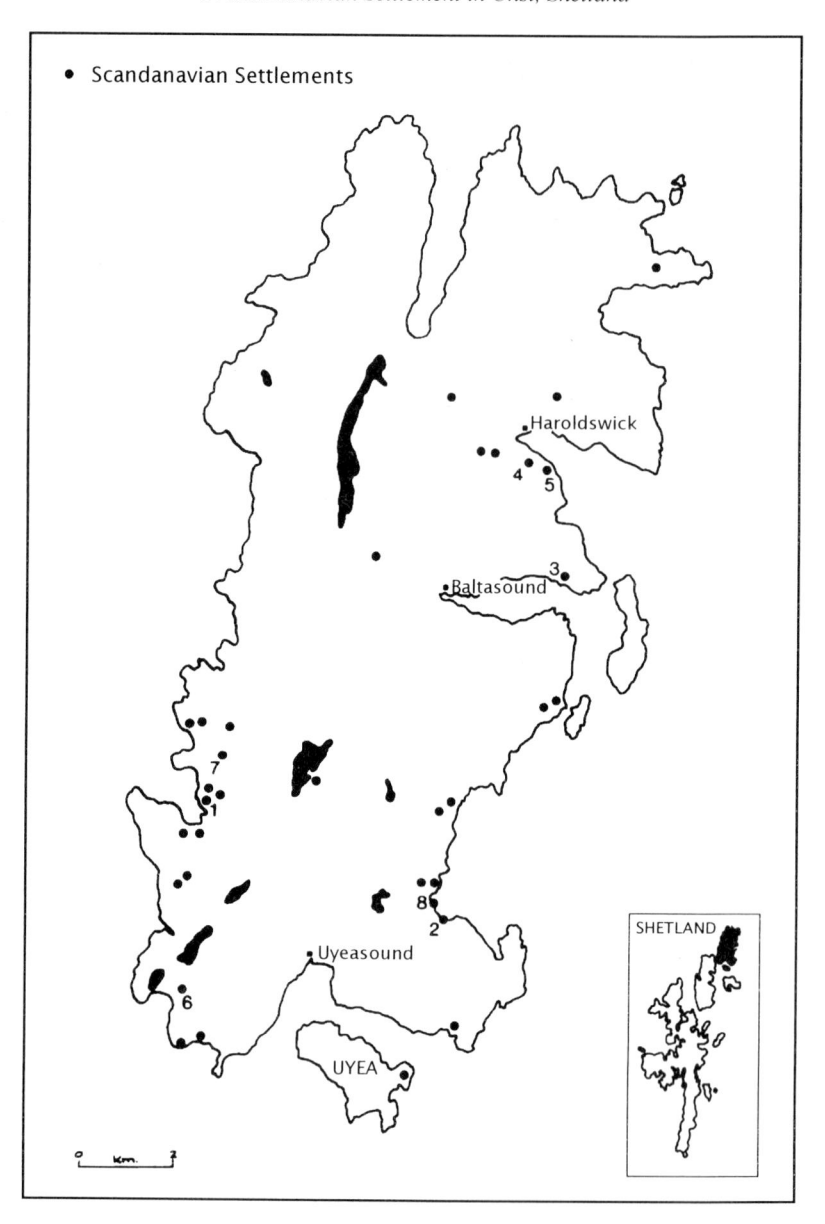

5.3: Distribution map of Norse sites in Unst
Sites referred to in the text: 1. Underhoull; 2. Sandwick-South; 3. Hamar; 4. Gardie I (Brookpoint); 5. Gardie II (Soterberg); 6. Setters (Wadbister); 7. Stoora Tofta, Westings; 8. Sandwick-North
(Map: J. Renny)

5: Scandinavian Settlement in Unst, Shetland

5.4: Norse longhouse at Hamar
(Photo: S. Stummann Hansen)

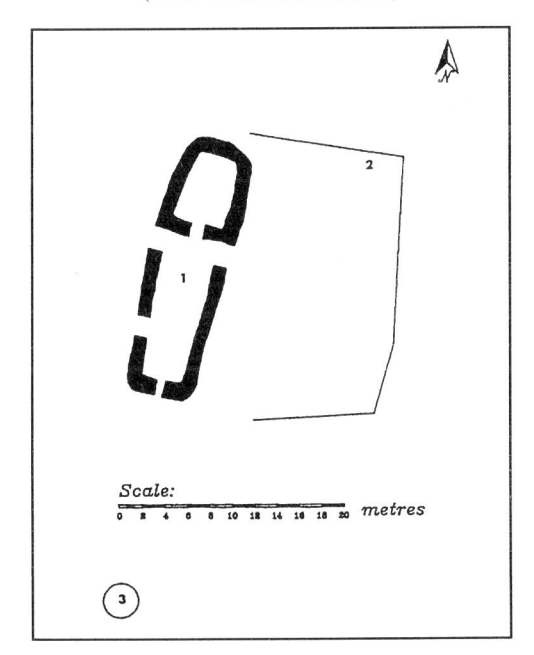

5.5: Plan of Norse site at Hamar
1. house structure; 2. stone wall around infield or farmyard;
3. hole for collecting the effluent from the byre

in a preliminary manner. In the spring of 1995, a trial excavation was carried out at the site of Hamar, where a perfectly preserved Scandinavian longhouse is situated aligned down slope (Fig. 5.4). The structure is 23 metres in length (internal measurements) and has a maximum internal width in the middle of the house of *c.*5 m.

The structure has curved walls and rounded corners. The walls were *c.* 1 m in thickness. Prior to the trial excavation the structure was properly planned in detail and photo-documented. A trial trench 60 cm in width was cut across the upper part of the structure. In this trench it could be documented that there had been *c.*80 cm wide benches along the walls. The floor-layer was located and it became clear that the bottom of it originally had been about 75 cm below the surface of the surrounding top-soil. In other words, the evidence of the small trial excavation indicates that the house had a sunken floor. The floor-layer was not touched but in the top-soil a single sherd of a steatite bowl was found, thus confirming the dating of the structure to the Scandinavian period.

After the trial excavation was finished, a full recording and mapping of other structures connected with the house took place. This included the remains of what is supposed to be a wall around the farmyard and a circular feature south of the lower-lying gable that may have been a hole for collecting the effluent from the byre (Fig. 5.5). There is some indication of another oblong building parallel to the house and attached to the inner side of the wall surrounding the farmyard.

At Brookpoint near Gardie (Gardie I), at the southern side of Haroldswick, another site was trial-excavated in 1995 (Fig. 5.6). This structure was also aligned down slope. Its internal measurements were 14 m in length and a maximum width of 4.5 m. It had curved walls and – like Hamar – an opening for the drain from the byre in the middle of the lower-lying gable end. It has had opposite entrances in both the eastern and western walls in the northern part of the house. The stone-built walls seem to have been approximately 1 m in thickness. Just north of the entrances, as at Hamar, a dividing wall can be seen. There are remains of an addition near the entrance in the eastern side-wall. North of the lower-lying gable is a circular feature that may have been a hole for collecting the effluent from the byre, while attached to the north-western corner of the building there are well-preserved remains of an enclosure. Finally, there are well-preserved traces of a stone-built infield fence, which is contemporary with the building. The only purpose of the small trial-excavation was to locate the floor-layer. No finds turned up, which is understandable as the floor-layer itself was not touched.

5: Scandinavian Settlement in Unst, Shetland

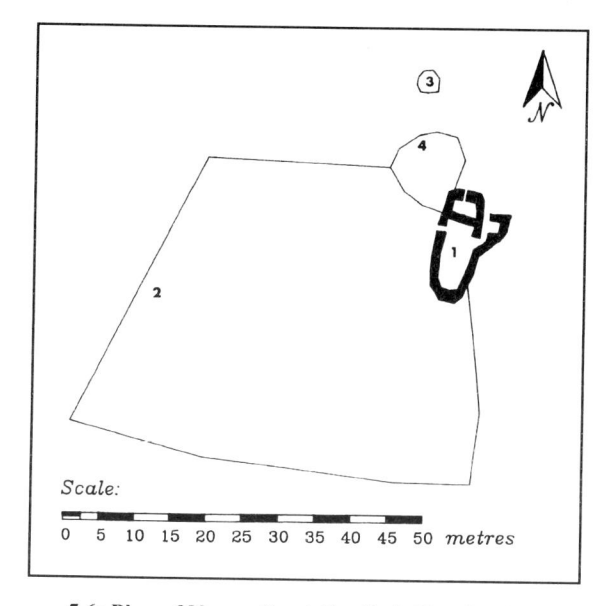

5.6: Plan of Norse site at Gardie I (Brookpoint)
1. house structure; 2. stone wall around infield or farmyard;
3. hole for collecting the outflow from the byre; 4. stone-walled enclosure

At Soterberg, close to Gardie I, another site was mapped. This multiperiod site, called Gardie II, contains remains of Norse structures very similar to those at Gardie I. Among other things, there is an identical enclosure wall also at the north-western corner of this building.

The site was the subject of a small local excavation in the 1970s and produced finds of Iron Age as well as Norse character as, for example, a spindle-whorl of steatite.

At Setters near Wadbister – or Belmont, which is the present-day name – Peter Moar had previously recorded a well-preserved Norse longhouse aligned down slope and of approximately the same size as that at Hamar. As at Hamar there is evidence that another parallel house might be present just next to it.[2]

Finally, the site Stoora Toft near Westing on the west coast of Unst should be mentioned. This building, which was located by Peter Moar, is of approximately the same size as Hamar and Setters and is like them aligned down. There are remains of attached outhouses close to the longhouse itself.

With this evidence we are now a little closer to the concept of a

Scandinavian farm in Shetland in the Viking and Early Medieval period. Thus it seems that the longhouses are typically aligned down slope. They have curved walls and rounded corners. The walls are generally at least 1 m in thickness. The houses may have sunken floors with benches along the inner walls in the upper part of the houses. The upper part of the house has probably been the sleeping room, while the central part has been what could be termed the living room with the fire-place. The lower-lying end has been the byre with a drain running out through an opening in the gable. Outside and below the byre end there often is a hole for collecting the outflow from the drain. These holes today appear as rounded swampy areas with a diameter of 2–3 metres. An enclosure wall in some cases seems to have been attached to the left corner of the byre end. This enclosure may have been used for the cattle. Also an earthern or stone-built fence has surrounded the farmyard. Parallel to the dwelling-house there may be another building, which has functioned as an outhouse. Where remains of an infield fence have been found, they seem to indicate that the house was attached to it.

It is worth noting that several of these structures are only preserved because they had been situated in areas which have not been subject to cultivation since the medieval or Late Norse period – thereby indicating a larger population and heavier pressure on the arable land in the Viking period.

With this survey, the previously recorded numerous preserved chapel-sites in Unst (Cant 1976, 1996; Lowe 1988) have now been matched by an identical abundance of Norse farmsteads. It is worth noting that Norse long-houses have been recorded beside medieval chapel-sites in at least two locations, Framgord and Colvedale, thereby demonstrating the potential for highlighting the relationship between church and farm in the Scandinavian period.

Excavations at Sandwick-North

It has already been mentioned that in the late 1970s Gerald F. Bigelow excavated a well preserved twelfth- to fourteenth-century Norse farmstead at the southern end of the sandy beach at Sandwick in Unst (Bigelow 1985, 1987, 1989, 1992). However, already at that time another site was known to exist at the northern end of the beach. This site, called Sandwick-North, was first identified by the Royal Commission on Ancient and Historical Monuments in Scotland (*RCAHMS* 1946, 139–40), but it was first confirmed as Viking or Norse in 1980, when a small trial excavation at the then heavily eroded site was conducted (Fig. 5.7). This trial excavation was referred to in a small article in *Discovery & Excavation in Scotland* (Bigelow, Butler &

5: Scandinavian Settlement in Unst, Shetland

McGovern 1980, 26).

During the present survey in Unst in the autumn of 1994 it became evident that the site, being situated within the current springtide zone, was apparently now eroding rapidly and the remains would probably soon be completely lost to the sea. Salvage recording and excavation were therefore desirable in order to try to date the site and record it as fully as possible. An excavation project was set up and excavation took place in 1995.[3]

5.7: View over the Sandwick-North site during excavation
Note the indication of Sandwick (far end of the beach) and Sandwick-North (right-hand arrow)
(Photo: S. Stummann Hansen/Shetland Amenity Trust)

The site appeared as a heap of stones with very little evidence of structures. Parts of walls were, however, to be seen in the southern as well as the northern part of the site. It was already obvious at that stage that these fragments of walls could not be part of the same structure.

In total an area of approximately 450 m² was excavated. The eastern limit of that area was in general defined by the erosion line, which was very clear. East of the erosion line, i.e. between the preserved structures and the sea, the beach sand was packed with boulders and stones indicating the erosion and the earlier complete demolition of buildings in that area. Inside the excavation area the very fragmentary remains of several structures were recorded and excavated.

It is the impression that at least four phases of habitation are represented on the site, but due to the erosion in most cases so little was preserved of the individual buildings that hardly anything can be deduced about the lay-out and the size of the buildings, let alone their function. In the northern part, however, a well-paved stone-built entrance and path were preserved. This entrance and path led into the best-preserved structure, which obviously had been part of an originally more extensive complex on the site. The structure consisted of two rather small rooms orientated approximately north to south and separated by a stone-built wall. The entrance to both rooms was from the above-mentioned entrance or path, which seems to have passed right through the complex with the two rooms. The two rooms seem to have been separated from one another by a stone-built wall. Before we can establish the function of these rooms, we will have to await further examination of finds, samples etc. They do not, however, seem to have been dwelling-rooms, as no fire-place was found in either of them.

Although the buildings on the site were only fragmentarily preserved, a large number of finds were uncovered during the excavation. The assemblage of artefacts was dominated by items of various types of stone materials, especially steatite and schist. They included large numbers of sherds from vessels (cooking pots, saucepans, cups, etc.), line- and net-sinkers, gaming-boards, spindle-whorls, baking-plates and whetstones. Other objects worthy of mention include bone-combs (Fig. 5.8), bone-pins and a beautifully preserved bronze-pin of Hiberno–Norse type.

All of these objects are typical of the Viking and Medieval Scandinavian emigrant culture in the North Sea and North Atlantic region. Most of them have their origin in Scandinavian culture, but it is difficult at the present time to say whether they were produced there and subsequently exported to the emigrant communities or if they were produced locally. While whetstones of schist, baking-plates of steatite and bone-combs may have been imported from Scandinavia, vessels and other objects of steatite would have been produced locally. Several steatite quarries are known in Shetland (Ritchie 1984), for example that excavated at Clibberswick in Unst (Clibber= Scandinavian [and Shetland] *kleber*, which means 'soap-stone' or 'steatite').

Often the lay-out and construction of buildings provide a good indication of their date. Curved walls in long-houses generally indicate a date to the period AD 800–1200, while straight walls seem to be a feature that gradually becomes typical in the following Late Norse or Medieval period.

The remains of the buildings preserved at Sandwick-North give the immediate overall impression of being – as already suggested by Bigelow in 1980 – from the Late Norse period, but as the structures were very

5: *Scandinavian Settlement in Unst, Shetland*

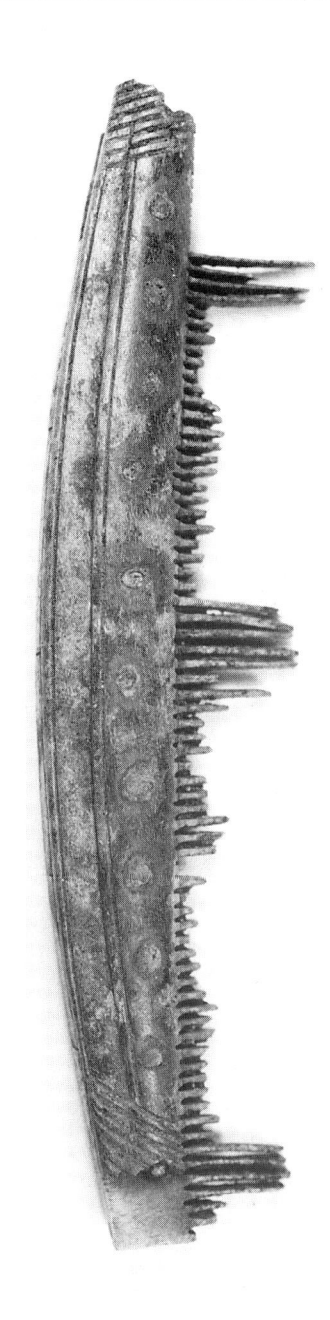

5.8: Ornamented bone-comb with copper rivets from the Sandwick-North site. Length 12.3 cm
(Photo: Shetland Museum)

fragmentarily preserved, they cannot provide us with a more exact date within the period. It is therefore necessary to turn to the objects found during the excavation in order to establish a closer date for the site.

The artefact assemblage is rather homogenous. The steatite sherds from the site represent vessel-types (mostly circular and oval) which are typical for the Viking as well as the Late Norse period. Baking-plates, however, are not represented in Viking contexts in Norway or the Norse emigrant communities in the North Atlantic and must be regarded as an artefact type from the Late Norse period. In Norway, the manufacture of baking-plates seems to start about 1100 AD (Weber 1984 and forthcoming). The combs from Sandwick-North also indicate a Late Norse date (eleventh to thirteenth century).

Finally, mention should be made of the bronze pin of Hiberno–Norse (Irish–Scandinavian) type. This type of pin, which is characteristic for the Scandinavian emigrant communities in Ireland and Scotland and uncommon in the emigrants' Scandinavian homelands, is represented in stratified layers in the Scandinavian settlement in Dublin but the assemblage there has not yet been fully analysed (Fanning 1994, 1ff). The pin is, however, definitely of Late Norse date and an eleventh-century date is suggested. An almost identical parallel to the pin from Sandwick-North was found years ago while excavating a Norse building at the Brough of Birsay in Orkney (Curle 1982, 62, Ill. 39:423).

An appropriate date for the settlement at Sandwick-North consequently must be eleventh to thirteenth century. Regarding the interrelationship between the Sandwick site, excavated by Bigelow in the late 1970s, at the southern end of the beach, and Sandwick-North, it should be mentioned that hardly any Norse pottery was found at Sandwick-North. Fragments of square-sided vessels similar to the one published by Bigelow from the Sandwick site were found in the top-layer of a midden, which is interpreted as the latest structure of all at Sandwick-North. In other words, there seems to be only a slight overlap or none at all between the two sites. This question cannot, however, be satisfactorily answered until a more detailed study of the records and the finds has been completed.

These thoughts about the excavation at Sandwick-North must necessarily be of a preliminary nature, as the excavation was only finished a few months ago. Ahead of us lies the post-excavation work, which includes sorting out the relationship between the different recorded structures, examination of objects and samples and determination as to species of the thousands of animal bones (especially fish).

Scandinavian Unst and Scandinavia

In the Viking and medieval period the North Sea region became a Scandinavian sphere of interest. The Viking expansion and emigration in the West resulted in the establishment of Scandinavian emigrant communities in England, Ireland, Scotland, the Western and Northern Isles and further out in the North Atlantic in the Faroe Islands, Iceland and Greenland. Ultimately there came the attempts to settle in North America around the year 1000.

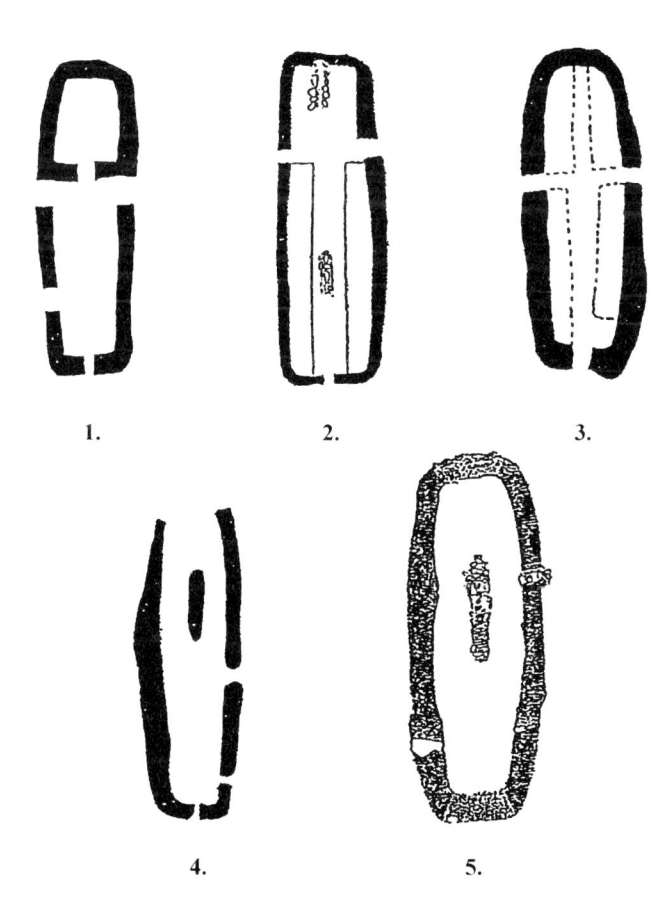

5.9: Norse longhouses of the ninth and tenth century
1. Hamar (House 1), Shetland; 2. Jarlshof (House 1), Shetland; 3. Birsay (Site C), Orkney; 4. Toftanes (House II), Faroe Islands; 5. Kvívík (Hóllin), Faroe Islands
(Drawing: S. Stummann Hansen)

Wherever the Scandinavian settlers went, they brought with them their architecture in the form of the classic Viking longhouse. They had a very strong concept of what a 'house' was and what 'home' was. The architecture had an almost symbolic character as the mobile Viking emigrants could travel everywhere in the North Sea and North Atlantic region and still feel at home (Fig. 5.9). They were virtually travelling in a Scandinavian world (Stoklund 1984; Stummann Hansen & Larsen forthcoming). The identical lay-out of house-structures all over this Scandinavian world offers us some promising possibilities to compare evidence from the rural settlements in Scandinavia with the rural settlements from, for instance, Shetland.

Unfortunately, there are rather few settlement sites from the Viking period excavated in Western Norway, and therefore, although rural settlements from the Viking and Early Medieval period here are only represented by postholes today, the best comparative material is actually found in Southern Scandinavia, and especially Denmark. When Johannes Brøndsted (1890–1965) published his second and revised edition of *Danmarks Oldtid* in 1960, only three locations with house-structures from that period were known (Brøndsted 1960, 371). This picture, however, changed rapidly in the 1970s when C.J. Becker of the University of Copenhagen initiated a large-scale project, funded by the State Research Council for Humanities in Denmark, on Viking Age settlements in Jutland (Becker 1980). Under this project, which was conducted during the years 1974–76, four localities were nearly completely excavated thus producing hundreds of house-structures from the Viking and early medieval period (Bender Jørgensen & Skov 1980; Bender Jørgensen & Eriksen 1995; Hvass 1980; Nielsen 1980; Stoumann 1980). The project demonstrated how to locate and excavate on a large-scale sites like that, and since that time hundreds of house-structures have been excavated all over Denmark and recently also in southern Sweden.

In Denmark the progressive stages in the development of the Scandinavian long-house throughout the Viking and early medieval period have gradually become very well-dated, especially through dendrochronology (Hvass 1993, 187–94; Bender Jørgensen & Eriksen 1995, 17–21) (Fig. 5.10). This provides us with a very good basis for dating relict house-structures in Shetland and everywhere in the Scandinavian emigrant communities. The advantages in combining this with studies on the chronological development in the use of place-names is obvious. Furthermore, the settlements in Shetland may provide information on details that long ago were destroyed by cultivation in Scandinavia: for instance benches, infield-fences, etc.

5: *Scandinavian Settlement in Unst, Shetland*

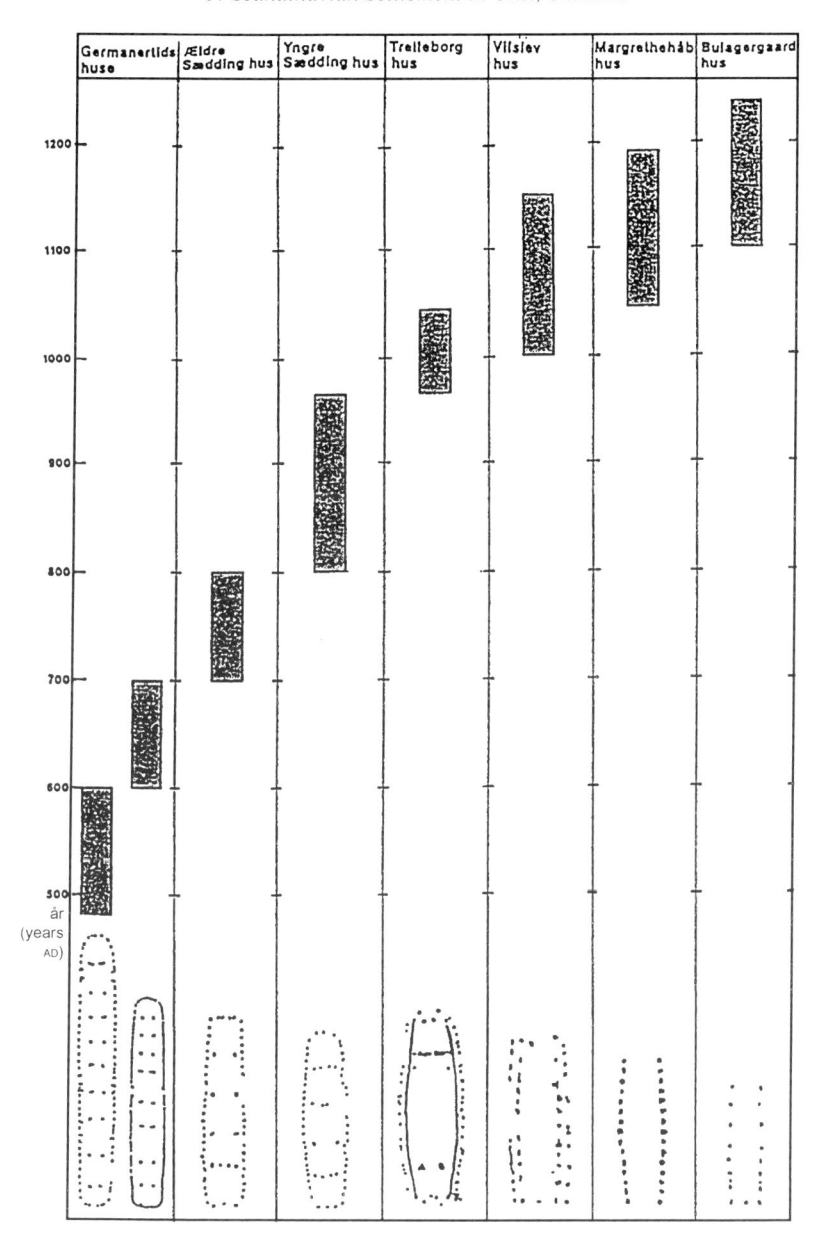

5.10: Typological development of Viking and Early Medieval architecture in Denmark
(After Bender Jørgensen & Eriksen 1995)

Perspectives for Future Research

It is in this perspective that the results of the pilot project in Unst should be seen. And it is in this perspective that a more detailed research design for a much more ambitious project on the Scandinavian settlement in Unst has been proposed by the author. The project, which hopefully will last for at least five years, so far includes recording of private collections, large-scale excavations of settlements and burial grounds, aerial survey, place-name studies (including field-work), pollen analysis, etc.[5] It is the hope that this joint project will continue in the spirit of Worsaae and his fellow colleagues in the British Isles and Ireland.

Acknowledgements

First of all I want to express my sincere thanks to the organisers for having invited me to speak at the most inspiring Day Conference on the Uses of Place-Names, at which the above chapter was first read as a paper. Further I want to express my gratitude to His Royal Highness Crown Prince Frederik's Foundation (Denmark) and the Shetland Amenity Trust, who provided the financial support for the survey in Unst. Also my thanks go to the Shetland Museum, Shetland Archives and the Department of Archaeology at the Royal Museum in Edinburgh for support and good co-operation. I am also grateful to Lindsay Potter, who helped us with the mapping of some of the Norse sites in Unst. Finally my thanks go to Professor Christopher D. Morris, University of Glasgow, for critical and constructive comments on this chapter.

Part II
Doreen Waugh

Steffen Stummann Hansen and I are just at the start of what I believe will be an exciting and productive period of co-operation. As he has shown, Unst is an island which is rich in Norse remains, both archaeological and linguistic, and our purpose is to work – each in his or her own field – to produce a mutually enlightening study of the archaeology and place-names of Unst and to consider comparative material from elsewhere in the Scandinavian homelands and colonies. The larger project proposed by Steffen Stummann Hansen is outlined above and it should be noted that, at this stage, research in the archaeological part of the larger survey is much in advance of that in any of the other disciplines involved.

When researching local names, I have the advantage of being a Shetlander and of having recent experience of conducting a study of the place-names of the village of Sand on the west side of Shetland, in the course of which I used a combination of documentary sources and oral tradition to build up a picture of settlement history as seen through the place-names (Waugh 1996, 242–54). Similar methods will be used in Unst to find out as much as possible about the Norse place-names which are either still extant in the vicinity of known archaeological sites throughout the island or which are recorded in documents relating to the area. Many farms have been abandoned in recent centuries but it is likely that place-names associated with these earlier farms will still be used by people living nearby and field work may uncover some previously unrecorded names, as proved to be the case in Sand. Documents from the medieval period in Unst are scant on the ground but there is some more recent material in the Shetland Archives which will be examined as part of the project.

At this stage (February 1996), therefore, before I have either investigated archival material in Lerwick and elsewhere or visited sites in Unst, I shall limit myself to some preliminary discussion of the names of places mentioned earlier in this chapter.

Some Scandinavian place-names

Steffen Stummann Hansen's preliminary trial excavation in the spring of 1995 took place at the site of Hamar, a place-name which derives from Old Norse *hamarr* 'a rocky outcrop on a hillside'. It is a name which occurs elsewhere in Shetland, with the most well-known example, other than the Hamar in Unst, being in Northmavine towards the north of mainland Shetland, where the rocky outcrop is used as a dry site for farm buildings while the adjacent patch of land is cultivated. The name, Hamar, tends to be northern and western in its distribution in Shetland, but that may have more to do with the geology of the islands than with any other factor.

It is not surprising that a topographical term should appear as the name of a Viking Age site because the Norse commonly used notable aspects of the landscape in naming, particularly in the early stages of settlement. Many of the villages of Shetland have topographical names, such as Voe – from *vágr* 'an inlet of the sea', Aith – from *eið* 'an isthmus', etc. When it became necessary to devise new names for neighbouring settlement, the Norse could then add to their generic, as in Littlehamar, Norhamar and so on. In the case of Hamar in Unst, the serpentine rock – an outcrop of which gave rise to the name – may have led to difficulties of cultivation and the site was abandoned. It is often difficult to distinguish between this particular Hamar and other examples of the name Hamar in Unst when these occur in published records, although, in some instances, helpful reference is made to *Hammer benorth* 1695 and *Hamber benorth the Voe* 1696 (Stewart 1987, 126).

Underhoull – the site which was investigated by Alan Small in 1963–65 –

is another example of a place-name based on the local topography. The name appears in written record at the end of the sixteenth century as *Underhoull* (Ballantyne & Smith 1994, 69) and it derives from Old Norse *undir* 'under or lower' and *hóll* 'a hill'. The reference is probably to the Norse settlement on the side of the hill where the excavation took place, although the Norse might have been inspired in their description by the neighbouring broch mound, which is a prominent feature of the landscape, but lower than the hill behind.

Sandwick, where archaeological work has also taken place, is like Hamar and Underhoull in being descriptive of the topography and it is a common place-name in Shetland and in Orkney. Jakob Jakobsen, for example, mentions three Shetland examples in Unst, Whalsay and Dunrossness (Jakobsen 1936; reprinted 1993). The name derives from Old Norse *sandr*, plus *vík*, 'sandy bay', and one can understand why the site would have been favoured in the Viking Age when relative ease of cultivation must have been a prime consideration in choosing a site, along with ease of access by boat although, judging from the map, the beach at Sandwick in Unst may be rather exposed. Stewart records the name as *Sandvik* in 1360 (Stewart 1987, 290) but, unfortunately, does not identify his source. It is certainly recorded in 1589 as *Sandwik* (Ballantyne & Smith 1994, 69).

Also mentioned in the earlier part of this article are Soterberg and Westing, both of which refer to land jutting out into the sea. Jakobsen records the former name as *Sotraberg* and derives it from *sátar-berg* (cf. Old Norse *setberg* 'angling-rock, a rock from which angling for small coalfish is carried on') (Jakobsen 1936; reprinted 1993, 25). Westing is the point of the long promontory of land (Old Norse *tangi*), stretching out towards the west, on which Underhoull is situated, and it is now invariably known as *Da* Westing ('The' Westing); in other words, the Norse compound was still understood when Norn was no longer the language of the area and it functioned as an appellative along with the English definite article in its Shetlandic form 'da'. It is interesting to note that the same happened in the case of *Da* Setters (Old Norse *sætr* 'a shieling'), which is the name preserved in oral tradition for the site near Belmont which is to be further investigated by the team of Danish archaeologists in the summer of 1996. Near Da Westing is another Norse longhouse with associated outbuildings, known as Stoora Toft, from Old Norse *stórr* 'large' and *topt, toft* 'a house site or piece of ground' and one wonders whether this is, in fact, a later description of the original Norse house once it had become ruinous.

The main villages of Unst – Uyeasound, Baltasound and Haroldswick – also fit into the pattern of naming by topography. Uyeasound and Baltasound both contain mention of islands lying offshore and to the channels of water running between these smaller islands and the larger island of Unst. Uyeasound is recorded by Jakob Jakobsen as *Øjasund* (Jakobsen 1936; reprinted 1993, 124), which derives from Old Norse *øy* 'an island' and *sund* 'a sound or strait'. Balta is the name of the island in the sound, but it contains a more specific reference to ownership of the island by a man called *Balti*, a name which is mentioned in Hákon Hákonsson's Saga as occurring in Shetland (Jakobsen 1936; reprinted 1993, 121). Haroldswick contains the Old Norse coastal term *vík*, already mentioned as occurring in Sandwick, but it is unusual among *vík*-names in that the specific is the personal name *Haraldr*. It is common for islands to be attributed to their Norse owners, as in Baltasound, but bays were much less likely to be thus attributed, which could suggest that this particular *Haraldr* was a man of some significance. Shetland folk etymology would certainly have it thus (see caption to Fig. 5.2).

In addition to place-names which are topographical in reference, there are, of course, numerous names in Unst containing Old Norse habitative generics such as *garðr* 'a yard or enclosure'. Those which are recorded on the map in the vicinity of Sandwick are Hannigarth, Housigarth, Houlligarth, Vatnsgarth and Smirgarth and it will be very interesting to see whether the archaeology of the area can provide clues as to their antiquity. *Garðr* – or more probably the related term *gerði* 'an enclosed patch of ground' – also occurs in the place-name Gardie, mentioned by Steffen Stummann Hansen as the name of a site on the southern side of Haroldswick (see above pp. 128f). Names derived from *garðr/gerði* – such as Framgord mentioned as the site of a medieval chapel – occur very frequently throughout the Northern Isles and suggest farming activities which used enclosure either as a means of parcelling out portions of land or as protection from grazing animals or, in some cases, the harsh climate, the adverse effects of which could be reduced by building protective walls. Like many other Norse names which must, originally, have been very specific in reference, Gardie now refers to an area rather than a precisely located place. Colvedale, the name of the other medieval chapel site mentioned, is also a name suggestive of farming activities, the specific in the name being Old Norse *kálfr* 'a calf'. Stewart records numerous examples of the name from 1360 onwards (Stewart 1987, 73) and it is recorded as *Colvadaill* in 1582 (Ballantyne & Smith 1994, 24).

Wadbister, a place adjacent to the archaeological site at Da Setters, is also a common Norse place-name in the Northern Isles and in the north of Scotland for a farm beside a stretch of water or loch (Old Norse *vatn*). The generic in the name is *bólstaðr* 'a farm or dwelling-place', commonly occurring as -*bister* in place-names in Shetland and Orkney. In this instance, the name Wadbister (Ballantyne & Smith 1994, 69) has been ousted by the imported name – Belmont – which was applied to the laird's house (built c.1777) situated at the head of the pier where the ferry from the neighbouring island of Yell terminates. The imported French name would have had flattering connotations of a social status higher than that to which ordinary Shetland crofter-fishermen could aspire. Wadbister, however, is preserved in the topographical name Ness of Wadbister. The eponymous *vatn* is now known as the Loch of Belmont and the bay into which the ferry sails is the Wick (Old Norse *vík*) of Belmont. Such linguistically hybrid compounds are common in Shetland and, of course, elsewhere. The process of replacing names with other names is on-going in Unst and, unfortunately, in this process some names are lost altogether because they are no longer in regular use. Some of the archaeological sites which are being investigated have, regrettably, lost their names, but it may prove possible to find a hint of these surviving in oral tradition, as in Sand.

Problems of name loss are not the only difficulties facing the researcher. Steffen Stummann Hansen begins his archaeological commentary on Scandinavian Shetland with mention of 'the most famous Viking site in Scotland... namely Jarlshof', but the name *Jarlshof* is far from Viking. It was invented by the novelist, Sir Walter Scott, who had a great interest in Old Norse culture and literature and who used the south of Shetland as the setting for his novel *The Pirate*. It is a tribute to the influential nature of Scott's work that the name Jarlshof, which he coined, is still with us today, perpetuated by the archaeological significance of the site where his novel was set.

Interest in Old Norse culture is also a hallmark of those engaged in the Unst project, but there will be no novelistic flights of fancy in the course of the project's implementation. Local research, both archival and oral, will help to flesh out the place-name picture, the very incomplete skeleton of which is sketched in this short commentary.

Notes

1. For more detailed references see Morris 1992.
2. As a co-operative venture between Shetland Amenity Trust and University of Copenhagen a large-scale excavation was initiated at the site in July 1996. The excavation was conducted by Anne-Christine Larsen, Copenhagen.
3. The excavation project was carried out by the author for Shetland Amenity Trust. Financial support for the rescue excavation was provided by Historic Scotland and Shetland Amenity Trust. The excavation was conducted during eight weeks in June–

July and September–October 1995.
4. The project is based on a co-operative venture between Shetland Amenity Trust and, through the author, the Institute of Archaeology and Ethnology, University of Copenhagen.
5. Ordnance Survey 1:25000 (OS) Sheet HP 40/50/60:642092.
6. Ordnance Survey 1:25000 (OS) Sheet HP 40/50/60:615025.
7. Ordnance Survey 1:25000 (OS) Sheet HP 51/61:630121.
8. Ordnance Survey 1:25000 (OS) Sheet HP 40/50/60:566009.

Bibliography

Ashmore, P., *Jarlshof – a walk through the past* (Historic Scotland, Edinburgh, 1993).
Ballantyne, J.H. and Smith, B., *Shetland Documents 1580–1611* (Lerwick, 1994).
Batey, C.E., 'Viking and Late Norse Caithness: The Archaeological Evidence', in *Proceedings of the Tenth Viking Congress*, Larkollen, Norway 1985, ed. J. Knirk. Universitetets Oldsaksamlings Skrifter Ny Rekke Nr. 9 (Oslo, 1987), 137–48.
Batey, C.E., 'Excavation of a Norse Horizontal Mill in Orkney', *Review of Scottish Culture*, 8 (1993), 20–8.
Batey, C.E. and Morris, C.D., 'Earl's Bu, Orphir, Orkney: Excavation of a Norse Horizontal Mill', in *Norse and later Settlement and Subsistence in the North Atlantic*, eds. C. Morris and J. Rackham (Glasgow, 1992), 33–42.
Becker, C.J., 'Viking-Age Settlements in Western and Central Jutland. Recent Excavations. Introductory Remarks', *Acta Archaeologica*, 50 (1979), 89–94 (Copenhagen, 1980).
Bender Jørgensen, L. and Eriksen P., 'Trabjerg. En vestjysk landsby fra vikinge-tiden', *Jysk Arkæologisk Selskabs Skrifter*, XXXI:1 (Højbjerg, 1995), English summary.
Bender Jørgensen, L. and Skov, T., 'Trabjerg. A Viking-age Settlement in North-west Jutland', *Acta Archaeologica*, 50 (1979), 119–36 (1980).
Bigelow, G.F., 'Sandwick, Unst and Late Norse Shetland Economy', in *Shetland Archaeology. New Work in Shetland in the 1970s*, ed. B. Smith (Lerwick, 1985), 95–127.
Bigelow, G.F., 'Domestic Architecture in medieval Shetland', *Review of Scottish Culture*, 3 (1987), 23–38.
Bigelow, G.F., 'Life in medieval Shetland: an archaeological perspective', *Hikuin*, 15 (Højbjerg, 1989), 183–92.
Bigelow, G.F., 'Issues and Prospects in Shetland Norse Archaeology', in *Norse and later Settlement and Subsistence in the North Atlantic*, eds. C. Morris and J. Rackham (Glasgow, 1992), 9–32.
Bigelow, G.F., Butler, S. and McGovern, T., 'Sandwick-North', *Discovery and Excavation in Scotland* (1980), 26.
Brøndsted, J., *Danmarks Oldtid III. Jernalderen*, 2nd edn (København, 1960).
Cant, R.G., *The Medieval Churches & Chapels of Shetland* (Shetland Archaeological and Historical Society, Lerwick, 1976).
Cant, R.G., 'The medieval church in Shetland: organisation and buildings', in *Shetland's Northern Links. Language and History*, ed. D. Waugh (Lerwick, 1996), 159–73.
Childe, V.G., 'Another Late Viking House at Freswick, Caithness', *Proceedings of the Society of Antiquaries of Scotland*, 77 (Edinburgh, 1943), 5–17.
Crawford, B.E., 'Papa Stour: Survival, Continuity and Change in one Shetland Island', in

The Northern and Western Isles in the Viking World, eds. A. Fenton and H. Pálsson (Edinburgh, 1984), 12–39.

Crawford, B.E., 'The Biggins, Papa Stour – a multi-disciplinary investigation', in *Shetland Archaeology. New Work in Shetland in the 1970s*, ed. B. Smith (Lerwick, 1985), 128–58.

Crawford, B.E., 'The excavation of a wooden building at The Biggings, Papa Stour, Shetland', in *Shetland's Northern Links. Language and History*, ed. D. Waugh (Lerwick, 1996), 136–58.

Crawford, I., 'Scot(?), Norseman and Gael', *Scottish Archaeological Forum*, 6 (1974), 1–16.

Crawford, I., 'War or Peace – Viking Colonisation in the Northern and Western Isles of Scotland', in Proceedings of the Eighth Viking Congress, *Medieval Scandinavia Supplements*, 2, eds. H. Bekker-Nielsen *et al.*, vol. ii (Odense, 1981), 259–69.

Crawford, I., *The West Highlands and Islands: a view of 50 Centuries: the Udal (North Uist) Evidence* (Cambridge, 1986).

Crawford, I. and Switzur, R., 'Sandscaping and C14: the Udal, N. Uist', *Antiquity*, 51 (1977), 124–36.

Curle, A.O., 'A Viking Settlement at Freswick, Caithness. Report on Excavations carried out in 1937 and 1938', *Proceedings of the Society of Antiquaries of Scotland*, 73 (1938–1939), 71–110 (Edinburgh, 1939).

Curle, A.O., 'Dwellings of the Viking Period. Jarlshof and Freswick', in 'Civilisation of the Viking Settlers in relation to their old and new countries', *Viking Antiquities*, VI, ed. H. Shetelig (Oslo, 1954), 9–63.

Curle, C.L., 'Pictish and Norse Finds from the Brough of Birsay 1934–1974', *Society of Antiquaries of Scotland Monograph Series*, 1 (Edinburgh, 1982).

Fanning, T., 'Viking Age Ringed Pins from Dublin', *Medieval Dublin Excavations 1962–1981* (Dublin, 1994), Ser. B., vol. 4.

Gelling, P.S., 'A Norse Homestead near Doarlish Cashen, Kirk Patrick, Isle of Man', *Journal of the Society for Medieval Archaeology*, 14 (1970), 74–82 (London, 1971).

Gelling, P.S., 'The Norse buildings at Skaill, Deerness, Orkney, and their immediate predecessor', in *The Northern and Western Isles in the Viking World*, eds. A. Fenton and H. Pálsson (Edinburgh, 1984), 12–39.

Hamilton, J.R.C, 'Excavations at Jarlshof, Shetland', *Archaeological Reports*, 1 (Ministry of Works, Edinburgh, 1956).

Hunter, J., Bond, J.M. and Smith, A.M., 'Some Aspects of Early Viking Settlement in Orkney', in *The Viking Age in Caithness, Orkney and the North Atlantic*, eds. C. Batey, J. Jesch and C.D. Morris (Edinburgh, 1993), 272–84.

Hvass, S., 'Vorbasse. The Viking-age Settlement at Vorbasse, Central Jutland', *Acta Archaeologica*, 50 (1979), 137–72 (Copenhagen, 1980).

Hvass, S., 'The Settlement', in *Digging in the Past. 25 years of Archaeology in Denmark*, eds. S. Hvass and B. Storgaard (Copenhagen, 1993), 187–94.

Jakobsen, J., *The Place-Names of Shetland* (London, 1936; reprinted Lerwick, 1993).

Kaland, S., 'Westnessutgravningerne på Rousay, Orknøyene', *Viking*, XXXVII (1973), 77–102.

Kaland, S., 'The Settlement of Western, Rousay', in *The Viking Age in Caithness, Orkney and the North Atlantic*, eds. C. Batey, J. Jesch and C.D. Morris (Edinburgh, 1993), 308–17.

Kaland, S., 'En vikingetidsgård og -gravplass på Orknøyene', in *Nordsjøen. Handel, religion og politikk*, J.F. Krøger and H-R. Naley (Karmøy, 1996), pp. 63–8.

Larsen, A-C., 'A Viking Age Farmstead at "Setters" in Unst', *Shetland Life*, 199 (Lerwick, 1997).

Lowe, C.E., 'Early Ecclesiastical Sites in the Northern Isles and Isle of Man: an Archaeological Field Survey' (unpublished Ph.D. thesis, Dept. of Archaeology, University of Durham, 1988).

MacLaren, A., 'A Norse House on Drimore Machair, South Uist', *Glasgow Archaeological Journal*, 3 (Glasgow, 1974), 9–18.

Morris, C.D., 'Viking Orkney. A Survey', in *The Prehistory of Orkney*, ed. C. Renfrew (Edinburgh, 1985), 210–42.

Morris, C.D., 'The chapel and enclosure on the Brough of Deerness, Orkney: survey and excavations, 1975–1977', *Proceedings of the Society of Antiquaries of Scotland*, 116 (Edinburgh, 1986), 302–74.

Morris, C.D., *The Birsay Bay Project, Volume 1. Coastal Sites beside the Brough Road, Birsay. Excavations 1976–1982* (University of Durham, Department of Archaeology Monograph Series number 1, Durham, 1989).

Morris, C.D., 'Viking and Late Norse Orkney. An Update and Bibliography', *Acta Archaeologica*, 62 (1991), 123–50 (Copenhagen, 1992).

Morris, C.D., *The Birsay Bay Project, Volume 2. Sites in Birsay Village and on the Brough of Birsay, Orkney* (Durham, 1996a).

Morris, C.D., 'The Norse Impact in the Northern Isles of Scotland', in *Nordsjøen. Handel, religion og politikk*, J.F. Krøger and H-R. Naley (Karmøy, 1996b), 69–83.

Morris, C.D., 'Church and Monastery in Orkney and Shetland. An Archaeological Perspective', in *Nordsjøen. Handel, religion og politikk*, J.F. Krøger and H-R. Naley (Karmøy, 1996c), 185–206.

Morris, C.D. et al., 'Skaill, Sandwick, Orkney: preliminary investigations of a mound-site near Skara Brae', *Glasgow Archaeological Journal*, 12 (Glasgow, 1985), 82–92.

Morris, C.D., Batey, C.E. and Rackham, D.J., 'Freswick Links, Caithness. Excavation and Survey of a Norse Settlement', *NABO Monograph*, No 1/*Highland Archaeology Monograph*, No 1 (Inverness and New York, 1995).

Morris, C.D., Batey, C.E. and Barrett, J.H., 'The Viking and Early Settlement Archaeological Research Project: Past, Present and Future', in 'Developments Around the Baltic and the North Sea in the Viking Age. Twelfth Viking Congress', eds. B. Ambrosiani and H. Clarke, *Birka Studies*, 3 (Stockholm, 1994), 144–58.

Nielsen, L.C., 'Omgård. A Settlement from the Late Iron Age and the Viking Period in West Jutland', *Acta Archaeologica*, 50 (1979), 173–208 (Copenhagen, 1980).

Parker Pearson, M. and Sharples, N., 'The Kilpheder Viking Age Settlement' (Excavation Proposal for 1996, unpublished, University of Sheffield, 1995).

Parker Pearson, M. and Webster, J., 'Bornish Mound 2 Viking Age settlement' (Interim report of the 1994 excavations, unpublished report, University of Sheffield, 1994).

Parker Pearson, M., Brennand, M. and Smith, H., '*Sithean Biorach, Cille Pheadair* "Fairy Point, Kilpheder". Excavations of a Norse Period Settlement on South Uist' (unpublished report, University of Sheffield, 1996).

RCAHMS, *Inventory of Monuments in the Outer Hebrides, Skye and the small Isles* (Edinburgh, 1928).

RCAHMS, *Twelfth Report with an Inventory of Ancient Monuments of Shetland and Orkney* (Edinburgh, 1946).

Ritchie, A., 'Excavations of Pictish and Viking-age farmsteads at Buckquoy, Orkney',

Proceedings of the Society of Antiquaries of Scotland, 108 (1976–77), 174–227 (Edinburgh, 1977).
Ritchie, A., *Viking Scotland* (Edinburgh, 1993).
Ritchie, P. Roy, 'Soapstone Quarrying in Viking Lands', in *The Northern and Western Isles in the Viking World*, eds. A. Fenton and H. Pálsson (Edinburgh, 1984), 59–84.
Roussell, Aa., *Norse buildings customs in the Scottish Isles* (Copenhagen/London, 1934).
Sharples, N., 'The Iron Age and Norse settlement at Bornish, South Uist: an interim report on the 1996 excavations', *Cardiff Studies in Archaeology, Specialist Report Number 1* (1996).
Sharples, N., Webster, J. and Parker Pearson, M., 'The Viking Age Settlement at Bornish, South Uist' (Interim Report of the 1995 Excavations, unpublished report, University of Sheffield, 1995).
Shetelig, H. (ed.), *Viking Antiquities in Great Britain and Ireland I–VI* (Oslo, 1940–54).
Small, A., 'Excavations at Underhoull, Unst', *Proceedings of the Society of Antiquaries of Scotland*, 98 (1964–1966), 225–45 (Edinburgh, 1967).
Stewart, J., *Shetland Place-Names* (Lerwick, 1987).
Stoklund, B., 'Building Traditions in the Northern World', in *The Northern and Western Isles in the Viking World. Survival, Continuity and Change*, eds. A. Fenton and H. Pálsson (Edinburgh, 1984), 96–115.
Stoumann, I., 'Sædding. A Viking-age Village near Esbjerg', *Acta Archaeologica*, 50 (1979), 95–118 (Copenhagen, 1980).
Stummann Hansen, S., 'Aspects of Viking-Age Society in Shetland and the Faroe Islands', in *Shetland's Northern Links: Language and History. Proceedings of the 21st annual conference of the Scottish Society for Northern Studies, Shetland July 1993*, ed. D. Waugh (Lerwick, 1996a), 119–37.
Stummann Hansen, S., 'Færøernes ældste historie – set i et arkæologisk perspektiv', in *Kontakter over Nordsjøen fra Merowingertid til slaget ved Hafrsfjord*, eds. F. Krøger and H-R. Naley (Karmøy, 1996b), 41–62.
Stummann Hansen, S., 'A Norse farmstead at Sandwick-North, Unst', *The New Shetlander*, 195 (Lerwick, 1996c), 28–9.
Stummann Hansen, S., 'Scandinavian Settlement in Shetland – its chronological and regional contexts', in *Proceedings of the Vikings in the West Conference* (Copenhagen, forthcoming).
Stummann Hansen, S. and Larsen, A-C., 'Viking Ireland and the Scandinavian Colonies in the North Atlantic', in *Viking Ireland*, ed. T. Damgaard-Sørensen (Roskilde, forthcoming).
Waugh, D. (ed.), *Shetland's Northern Links: Language and History* (Lerwick, 1996).
Weber, B., 'I Hardanger er Qverneberg og Helleberg...og Hellerne, det er tyndhugne Steene, bruger man til at bage det tynde fladbrød paa...', *Viking*, XLVII (1983), 149–60 (Oslo, 1984).
Weber, B., *The Biggins, Papa Stour, Shetland. Bakestones from the Excavations 1977–82* (forthcoming).
Worsaae, J.J.A., *Minder om de Danske og Nordmændene i England, Skotland og Irland* (København, 1851).
Worsaae, J.J.A., *An Account of the Danes and Norwegians in England, Scotland and Ireland* (London, 1852).

6

Place-names and Literature: Evidence from the Gaelic Ballads

Donald E. Meek

Introduction

Place-names have a very close connection with literature in all languages, and Gaelic is no exception. By 'literature' in this context I mean 'creative literature', in which the mind of a composer fashions a story or a poem. Furthermore, the concept of 'literature' which I use here embraces not only 'written literature' but also 'oral literature'; that is to say, stories and songs which were composed orally and transmitted orally, perhaps for many centuries, before being recorded in a manuscript. As is well known, there are considerable amounts of Gaelic 'literature' which were not designed for reading; many forms of Gaelic narrative were created to be recited or sung in an oral context, at least in the first instance. However, we must be careful not to overdo our emphasis on 'oral literature'; there are samples of Gaelic literature which did indeed begin life on a manuscript page, and passed into oral transmission at a later stage. Oral and written forms of literature co-existed in the Gaelic world throughout the Middle Ages. The Gaelic ballads which I wish to use as primary evidence in this chapter often functioned in an oral context, but they are also attested in manuscripts, in Ireland and Scotland, from the Middle Ages to the present day (Meek 1991).

I have chosen the Gaelic ballads as the primary evidence of this chapter for two reasons. First, I am particularly familiar with this genre of literature, having made an extensive study of one particular body of verse (Meek 1982); and second, the Gaelic ballad genre and the wider narrative cycle to which most of it belongs, the *Fianaigheacht* ('*Fian*-lore') as it is conveniently known in Ireland, are very closely connected with the landscape, and especially the toponymy, of both Scotland and Ireland. Indeed, I would reckon that, of all the various narrative genres in Gaelic, the ballads have the closest interactive relationship with place-names. The prose narratives which also belong to the tradition of *Fian*-lore have significant place-name components, but the ballads appear to preserve by far the greater toponymic interest, possibly because they may derive originally from a form of verse which commemorated and explained place-name lore. They have very close connections with what is called *dindshenchas* ('the lore of famous places'), a form of popular etymologising which is found in prose and verse forms in the Middle Ages (Murphy 1961; Ó hÓgáin 1988).

The *Fian*-lore records the many adventures of the warriors associated with Fionn mac Cumhaill, Finn mac Cool, traditionally believed to have lived in the third century AD. The names of the immediate members of his family are well-known, partly through the influence of Macpherson's 'Ossian' in the mid-eighteenth century: Ossian, his son, is the poet to whom many ballads (and also James Macpherson's 'translations') were ascribed in the course of the centuries. The ballads describe the adventures of Fionn's warriors, who belong to *fian*-bands. The root of the word *fian* is cognate with that of the Latin verb, *venare*, to hunt, thus denoting a group of young men whose lifestyle revolved around the hunt. Whatever difficulties there may be in establishing the historicity of Fionn mac Cumhaill (generally considered to have been a divine figure), there is no question about the historicity of *fian*-bands. It would seem that they were a well recognised institution in early Gaelic society, and that participation in a *fian*-band was one of the ways in which the young men of Ireland (and, we may presume, of Gaelic Scotland too) burnt off their excess energy before taking up a more settled life within the norms of the *tuath*, the primary political unit of early Gaelic society (McCone 1986). When they were not hunting, the *Fiana* were engaged in warrior pursuits of other kinds, including fighting Fionn's enemies, defending the land against invasion, and encountering beings from the Otherworld. Their adventures took place mainly in the open air, and for that reason the narratives of their activities in Ireland and Scotland are often studded with place-names marking their routes and the locations of their adventures.

The majority of Gaelic ballads belong to the wider body of *Fian*-lore, represented by two poems which are discussed in this chapter, namely *Laoidh Dhiarmaid* ('The Lay of Diarmaid') and 'Caoilte and the Creatures', but there are examples of very popular ballads which belong to other narrative cycles. A few are linked to the Ulster Cycle of tales, telling of the deeds of Cú Chulainn and other heroes from the old province of Ulster. One famous ballad, *Laoidh Fhraoich* ('The Lay of Fraoch'), which we shall discuss extensively, appears to have its narratological roots in Connacht, but,

6: Place-names and Literature

because of Connacht's connection with the Ulster Cycle, it is usually tagged on to that cycle. Ballads are also associated with the Arthurian Cycle of tales, which tell of the tests and quests of Arthur and his warriors (Gillies 1981).

The majority of the Gaelic ballads which are known to us today were composed in the later Middle Ages, from about the twelfth to the sixteenth century. They were composed in forms of syllabic metres which did not aspire to the degree of precision and strictness in the use of rhyme, alliteration, etc. which characterises bardic verse. Their language was based on the classical dialect shared by Ireland and Gaelic Scotland, but their composers appear to have made significant concessions to vernacular forms of the Gaelic language. This helped to ensure the survival of the ballads beyond the classical period; many passed into general currency, and their texts were reshaped and adjusted gradually to come more closely into line with the conventions of the spoken, modern languages (Gaelic and Irish) (Meek 1987). Thus, 'The Lay of Fraoch', which was probably composed in the fifteenth century, was still being sung in fragmentary forms in the Hebrides in the later twentieth century (MacInnes 1987).

The ballads were often transmitted orally, but they were also preserved in manuscripts compiled from the twelfth to the nineteenth century. The most important Scottish manuscript containing ballads is the Book of the Dean of Lismore, written in Perthshire between 1512 and 1542. All the texts which I am going to discuss in this chapter are found in that manuscript. The Book of the Dean is as complex as it is significant in the history of Gaelic literature; its scribes, James MacGregor, Dean of Lismore, and his brother Duncan, wrote Gaelic in a form of spelling based on that of Middle Scots, and, as a result, one has to 'crack the code' in the process of editing material from the manuscript (Meek 1990a). I am profoundly grateful that I am not engaged in code-breaking of that kind in this chapter; I am more than happy to be concerned only with the uses of place-names in the ballad texts. I shall simplify my remit still further by concentrating on poems known in Scotland.

Place-names can be used in a variety of ways by ballad composers, and I would like to offer an overview of some of these, employing some broad strokes of the brush, but also illustrating them in a more particular manner from the Gaelic ballad material.

1. Place-names and creativity

Place-names were an obvious stimulus to the creation of ballad texts. The difficulty lies in knowing the precise level, or significance, of the contribution of place-names in the creative process.

It does seem likely that a place-name, or a cluster of place-names, could form the starting-point of a distinctive version of a story or poem. A story thus created, or perhaps re-created, would seek to give meaning to, or derive some meaning from, a toponym or set of toponyms, following the broad patterns of the genre of composition called *dindshenchas*, in which place-names and their significance are explained, often in terms of a story.

An outstanding example of a ballad which employs a set of place-names which may have contributed significantly to the morphology of the narrative is *Laoidh Fhraoich* ('The Lay of Fraoch') (Meek 1984). This poem, probably composed in the fifteenth century, features the death of a Connacht warrior, Fraoch. The agent of his death is Meadhbh, queen of Connacht, who is jealous of his love for her daughter, Fionnabhair. Meadhbh feigns an illness which can be cured only by means of rowan berries, which grow on an island in a loch. The rowan tree is protected by a monster. Fraoch has to go to fetch the berries, but when he obtains them, Meadhbh requests a whole branch taken from the root of the tree, and he has to go back to the island; as he swims towards the land for the second time, Fraoch is noticed by the monster, which attacks and kills him.

The story in the ballad is an international one, as Dr George Henderson has demonstrated; tales about water monsters which attack humans are well known (Campbell and Henderson 1981). Nevertheless, a distinctively Gaelic version of this motif has emerged in 'The Lay of Fraoch', and place-names have contributed to that distinctiveness. According to the ballad, the hero, Fraoch, is buried at a place called *Carn Fraoich*, which, for the purposes of the ballad, is taken to mean 'The Cairn of Fraoch'. The primary place-names in the story are evident in the opening two quatrains of the ballad in the Book of the Dean, restored here in a tentative form, pending my forthcoming edition (Meek 1982, 369–77):

> *Osnadh carad a Cluain Fraoich,*
> *osnadh laoich a gcasail chró;*
> *osnadh dhian ní tursach fear*
> *agas da n-guileann bean óg.*
>
> *Ag so shoir an carn fán bhfuil*
> *Fraoch mac Fiodhaigh an fhuilt mhaoith,*
> *fear a rinn buidheachas badhbh –*
> *is bhuaithe shloinntear Carn Fraoich.*

(The sigh of a friend comes from Cluain Fraoich, the sigh of a hero in a covering of blood; it is an intense sigh which saddens a man, and which causes a young woman to weep.

6: Place-names and Literature

>   Over here in the east is the cairn under which lies Fraoch son of
>   Fiodhach of the soft locks, a man who made scald-crows happy – Carn
>   Fraoich takes its name from him.)

The use of the place-name in this manner, pointing to the grave of the hero, maintains the veracity of a form of the story which results in the death of Fraoch. This is important because there was another version of the story, represented in the eighth-century Early Irish tale *Táin Bó Fraích* ('The Driving of Fráech's Cattle'), which claimed that Fraoch was resuscitated following his fight with the monster, and lived to participate in further adventures. The place-name *Carn Fraoich* fixes the tragic outcome of the story, thus demonstrating the importance of the use of a place-name to preserve a particular version of a tale (Meek 1984).

There is a very strong possibility that the place-name *Carn Fraoich* was 'recycled' by the composer of the Gaelic version of the ballad. Originally it may have meant nothing more than 'Cairn of Heather', since *fraoch* commonly means 'heather' in Gaelic and Irish. However, *fraoch* can also mean 'bristle', which has a secondary meaning of 'rage, anger', making it an appropriate name for a warrior. The possibilities of connecting a simple place-name with a well-named Gaelic warrior will thus be evident.

Because of Fraoch's connections with Connacht and the prominence of Meadhbh, queen of Connacht, in the text, it is natural to assume that the ballad was composed in that part of Ireland. Indeed, the region seems to furnish what may be the primary locus of the ballad, since the principal place-names of the story are all to be found in Co. Roscommon, within a few miles of one another. The *Carn Fraoich* of the ballad is perhaps to be equated primarily with Carnfree in Co. Roscommon. The Connacht location includes, in close proximity, other names which appear to correspond to places in the topography of the ballad; most notable is *Cluain Fraoich* ('Meadow of Heather'), mentioned in the opening line of the ballad and twice more in the text.

Loch Máigh, the loch in which the tragedy occurs, is probably to be equated with Loch Baah, near Castleplunket. It may have come to be associated with the narrative by (mis)construing the original form of the name as *Loch mBágha*, 'Loch of the Fighting(s)/Contest(s)'. Although it may be dangerous to maintain that the earliest version of the ballad was composed in Co. Roscommon, it is safe to say that the text of the ballad as we have it in the Book of the Dean of Lismore was apparently shaped in that part of Ireland (*ibid.*). (See Figures 6.1, 6.2.)

6: Place-names and Literature

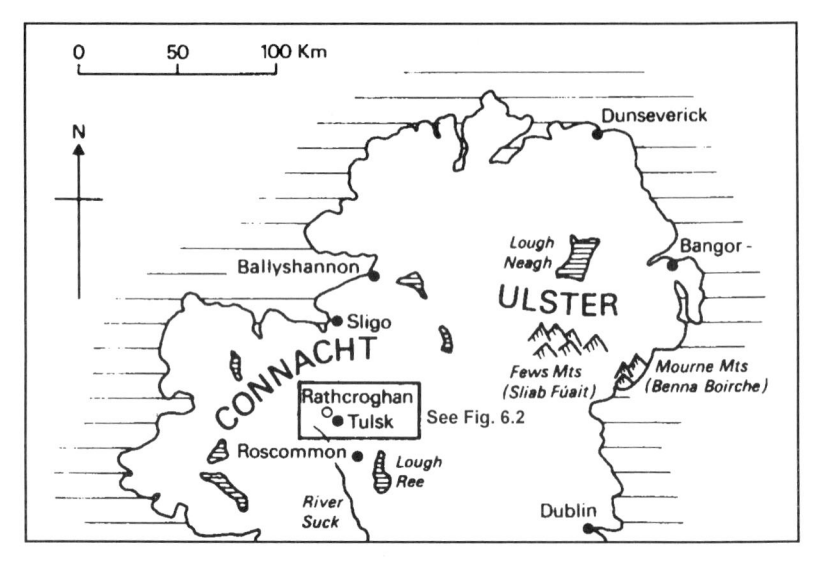

6.1: Irish locations associated with Fraoch traditions

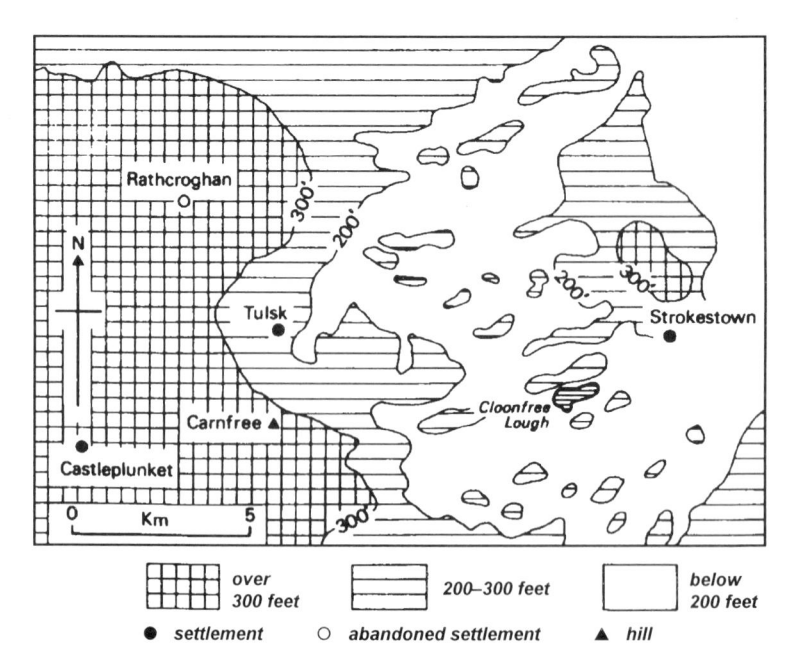

6.2: Place-names in Connacht connected with the *Laoidh Fhraoich* ('The Lay of Fraoch')

Whatever the region in which the ballad may have originated, place-names have played a significant part in creating a Gaelic version of an international tale and in creating our existing text. They have placed the tale in a convincing Gaelic context, although (as we shall see) the ballad could find a particular location in more than one district.

2. *Place-names and verisimilitude*

Almost indistinguishable from the role of place-names as a stimulus to creativity is their use to provide anchoring for the narrative of a poem. A place-name or set of place-names can be harnessed to define a domain for the event or the action in the narrative, thus imparting actuality or verisimilitude to the story. The place-names used in the construction of 'The Lay of Fraoch' may be said to have this role as one of their functions, though their primary function may have been creative, in the sense that they contributed to the emergence of a particular local version of an international story. For the imparting of verisimilitude, place-names did not need to have a creative role through etymological speculation; their main purpose was reinforcement of the storyline.

Place-names could be employed in two ways to create verisimilitude:

(1) Place-names could offer a 'local focus' for a story, and thus provide affirmation of the indigenous nature of the narrative. A place-name or set of place-names might be invested with a new significance associating it/them closely with a particular story, rather in the manner of *dindshenchas* ('the lore of famous places'). As a consequence, the story itself would gain an added depth and poignancy, if the scene of the action could be identified within a locality.

This process appears to have happened in the case of a particularly popular *Fian* ballad, *Laoidh Dhiarmaid* ('The Lay of Diarmaid'), telling how the warrior Diarmaid was killed when hunting a venomous boar on a mountain called *Beann Ghulbainn* (Meek 1990b). Again, this is an international theme; like Achilles, Diarmaid is vulnerable in only one part of his foot; in Achilles' case it was his heel, while with Diarmaid it was his sole. Having succeeded in killing the boar, Diarmaid appears to overcome both his enemy and his weak spot. Sadly, however, the boar's poisoned bristle penetrates his sole when (at Fionn's request) he attempts to measure its carcase against the grain of the bristle.

The place-name *Beann Ghulbainn* is an almost invariable part of the narrative of Diarmaid's death; a hill or mountain of this name is often conveniently to hand wherever the story occurs in the Gaelic world. The name means literally 'Mountain of Snout', *gulba* ('beak, snout') defining the geographical feature which is characteristic of such hills, namely the tapering snout or spur which projects from the hill. The profile of the main part of the hill resembles a plateau. This configuration can be found in Ireland, as in the celebrated Benbulben in Co. Sligo, which is directly associated with the death of Diarmaid (*ibid*. 338–9). (See Figures 6.3–4.) However, it appears to be particularly common in Scotland. In Scotland the primary locus of the Diarmaid story seems to be *Beann Ghulbainn* in Perthshire [Ben Gulapin, NO1072], a splendid example of a 'snouty' mountain at the head of Glenshee (see Figure 6.4b). The version of the poem in the Book of the Dean of Lismore begins as follows (*ibid*. 352–7):

> *Gleann Síodh an gleann so rém thaoibh*
> *i mbinn faoidh éan agas lon;*
> *minic rithidís an Fhéin*
> *air an t-srath so an déidh a gcon.*
>
> *An gleann so fá Bheann Ghulbainn ghuirm*
> *as h-áilde tulcha fa ghréin,*
> *níorbh annamh a shrotha gu dearg*
> *an déidh shealg o Fhionn na bhFéin.*

(This glen beside me is Glenshee, where blackbirds and other birds sing sweetly; often would the *Fian* run along this glen behind their hounds.
This glen below green Beann Ghulbainn, whose knolls are the fairest under the sun – not infrequently were its streams red after hunts had been held by Fionn of the *Fiana*.)

Although Glenshee may be the primary Scottish location of the ballad, there are a number of other 'snouty' mountains called *Beann Ghulbainn* throughout the Highlands and Islands; one [Ben Gullipen, NN5904] is found in the Trossachs near Loch Venachar, and another [Beinn Ghuilbin, NH8917] to the north of Aviemore (see Figs. 6.4c and 6.4d).

It is not at all clear why, in certain cases, such hills should have been connected with this story; there is no obvious etymological link between the hero and the mountain, as there is in the case of Fraoch and *Carn Fraoich*. Is it possible that the snout of the hill was somehow seen to resemble the snout of a boar, and that this prompted an automatic association between this type of hill and the story of Diarmaid's death? The geological profile of these

6: Place-names and Literature

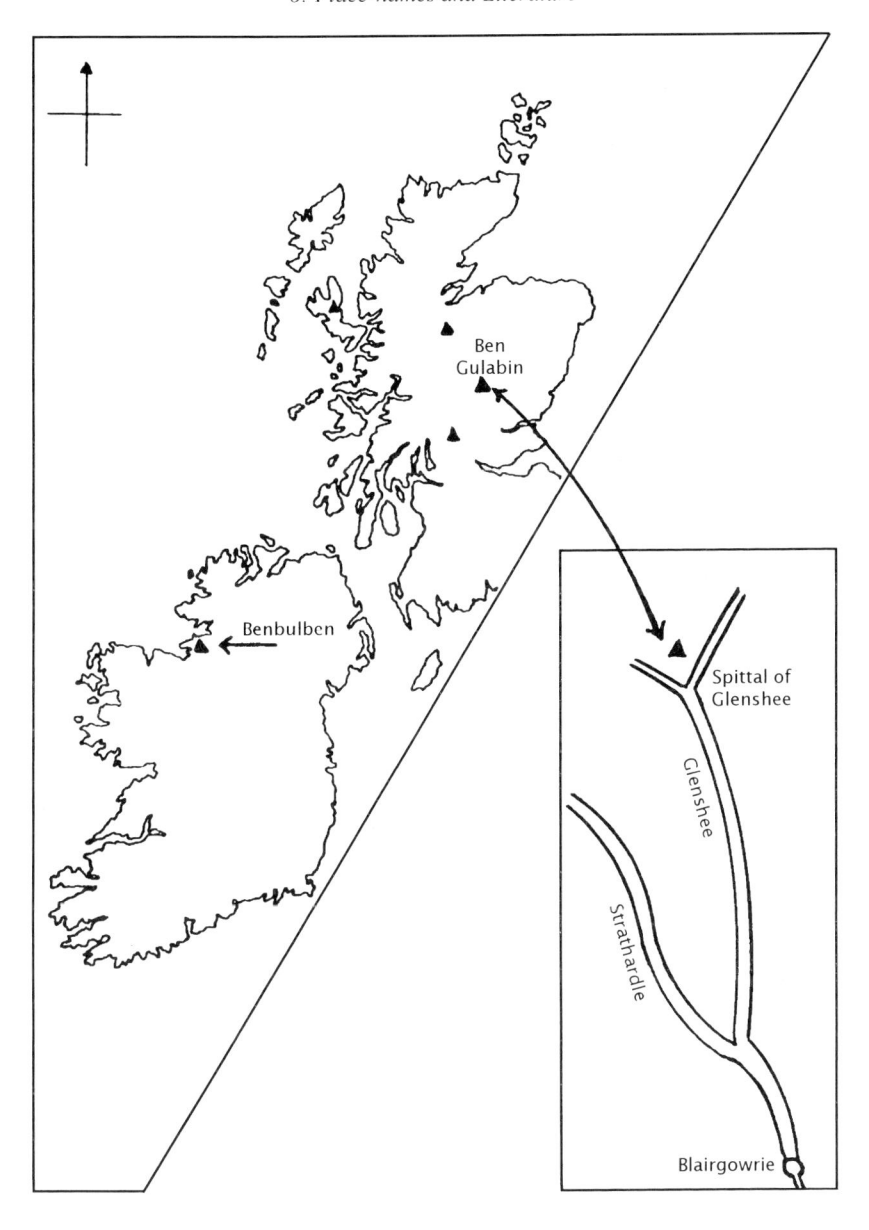

6.3: Locations of *Beann Ghulba(i)nn* in Ireland and Scotland

6: Place-names and Literature

6.4a: Benbulben, Co. Sligo
(Reproduced from P. MacCana, *Celtic Mythology*, London 1970, p. 103)

6.4b: *Beann Ghulbainn*, **Glenshee PER**
(Crown Copyright: Royal Commission on the Ancient and Historical Monuments of Scotland)

6: Place-names and Literature

6.4c: *Beinn Ghuilbin*, **north of Aviemore**
(Photo: Dr J. Grant)

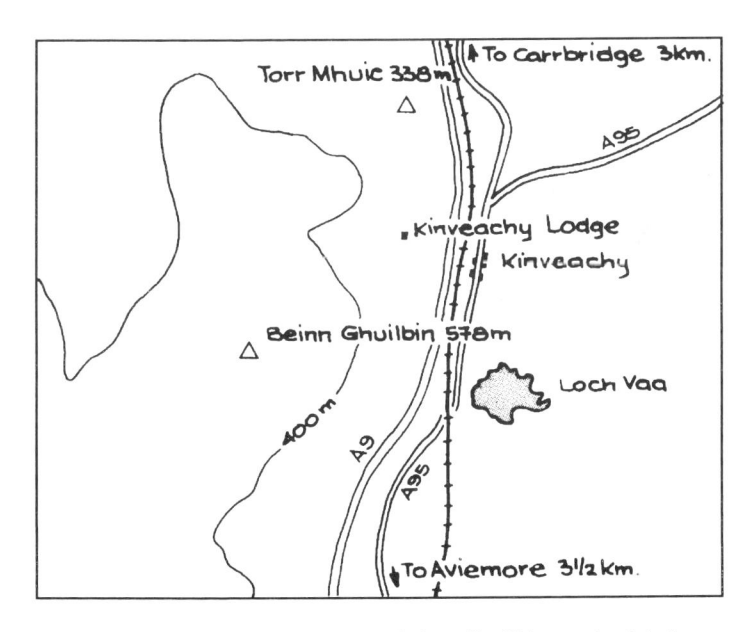

6.4d: Map showing exact position of *Beinn Ghuilbin* north of Aviemore
Note also the nearby Torr Mhuic (Map: J. Renny)

hills is often very striking; they rise dramatically from the surrounding countryside, and when seen from an oblique angle, they can sometimes recall the profile of crouching boars.

(2) Place-names could be used to confirm the action of the narrative, not only in terms of its location, but also in terms of the prowess of the participant(s). As we have seen, place-names like *Carn Fraoich* and *Beann Ghulbainn* were invested with a special significance by virtue of their association with a particular narrative, but such enhancement of the place-name was not always intended or necessary. A set, or catalogue, of place-names could be utilised without any attempt to impart a new significance to the place-names themselves, but merely to confirm the credibility of a narrative. Such catalogues were commonly used in ballads narrating quests, in which a warrior or warrior-band from the *Fian* had to participate in a significant journey of some kind (Meek 1986).

This is exemplified to a remarkable degree in a Gaelic ballad often called 'Caoilte and the Creatures', and found in the Book of the Dean of Lismore (Meek 1982, 240–57, 432–72). This poem tells how Fionn mac Cumhaill was taken prisoner by Cormac mac Airt, the legendary High King of Tara. He was to be released on condition that Caoilte, the fast runner of the *Fian*, would gather a pair of all the wild animals in Ireland, and bring them to Cormac at Tara. The main feature of the ballad text is a massive catalogue of place-names in Ireland, identifying the locations of the pairs of animals, and obviously the places where Caoilte caught them. Over fifty of these eighty or so place-names can be identified, and, when placed on a map, they show an interesting pattern which amounts to a rather chaotic circuit of Ireland. One is meant to admire the prowess of Caoilte, and not, of course, the cataloguing skills of the composer(s) of the sequence! (see Figure 6.5).

3. Place-names and mood-evocation

Place-names could be used not only to provide verisimilitude, but also to evoke a mood, as a kind of short-hand for a particular nuance or ambience which the composer felt was appropriate to the story. The evocative use of place-names has a long history in Gaelic literature.

One example of this can be given from the ballads. The name of *Eas Ruaidh* (Assaroe: 'The Waterfall of Ruadh') in Co. Donegal is strongly associated with the Otherworld. It would seem that it was a contact-point between the Otherworld and the world of flesh-and-blood warriors. It is associated with beauty and wisdom too. For these reasons, it features in the Gaelic ballads. Sometimes *Eas Ruaidh* is the starting point of adventures for the *Fian* warriors, particularly those involving the arrival of strangers of great beauty who seek the protection of the *Fiana*. It is also used in a shorthand manner to provide a bardic thumb-nail sketch of a warrior, like Diarmaid, who is described in the 'Lay of Diarmaid' as *seabhag súlghorm Easa Ruaidh*, 'the blue-eyed hawk of Assaroe' (Meek 1990b, 356, 361). This links Diarmaid with a mythological seam in his background. If the 'Lay of Diarmaid' was composed in Scotland (as I would contend!), the use of an Irish place-name would resonate in other ways too. It would underline the pan-Gaelic dimensions of Diarmaid, and also the pan-Gaelic nature of the wider ballad tradition. Scotland and Ireland shared the ballad genre, and Scottish composers would have known significant Irish names.

4. Place-names and general contextualisation

As the function of *Eas Ruaidh* would indicate, place-names played a part in the broader contextualisation of the tradition. The names of major centres of activity, relating to the dynamics of the wider cycle of tales, could be inserted naturally in the course of the narrative, in much the same manner as a story about King Arthur might be expected to mention Camelot at appropriate points in the text. Thus, the main centre associated with Fionn mac Cumhaill, namely *Almhain* or Allen in Co. Kildare, would be used in many texts merely as a reference-point for the story. At this level there were place-names which were in fairly common currency in Ireland and Scotland and likely to be known by most composers.

Place-names beyond the Gaelic world could also be utilised, if the narrative had an exotic touch. Such place-names were usually less specific in terms of their focus, and their associations were likely to shade into mythology. In ballads on Viking themes, for example, *Lochlann*, the popular name for 'Norway', occurs fairly frequently, and so also does *Baile na Beirbhe*, the Gaelic name for Bergen (Christiansen 1931). Similarly, storiological countries such as *Tír (Bhárr-)fo-Thuinn* ('Land [of Summits] below the Waves') could send emissaries to Ireland to meet the *Fiana*, and villains could come from such places as *Sorcha* ('Bright Place'). The *Fiana* could also travel on quests which took them to lands such as France, Germany and Spain, in the manner of crusaders. These lands are not necessarily to be identified with the modern countries of the same name. In interweaving 'real' and imaginary countries of this kind, some of the ballads share common ground with the Romantic tales (Bruford 1969, 21–2).

5. 'Transferable toponymy'

The ballads constituted a highly mobile form of literature, passed through the

6: Place-names and Literature

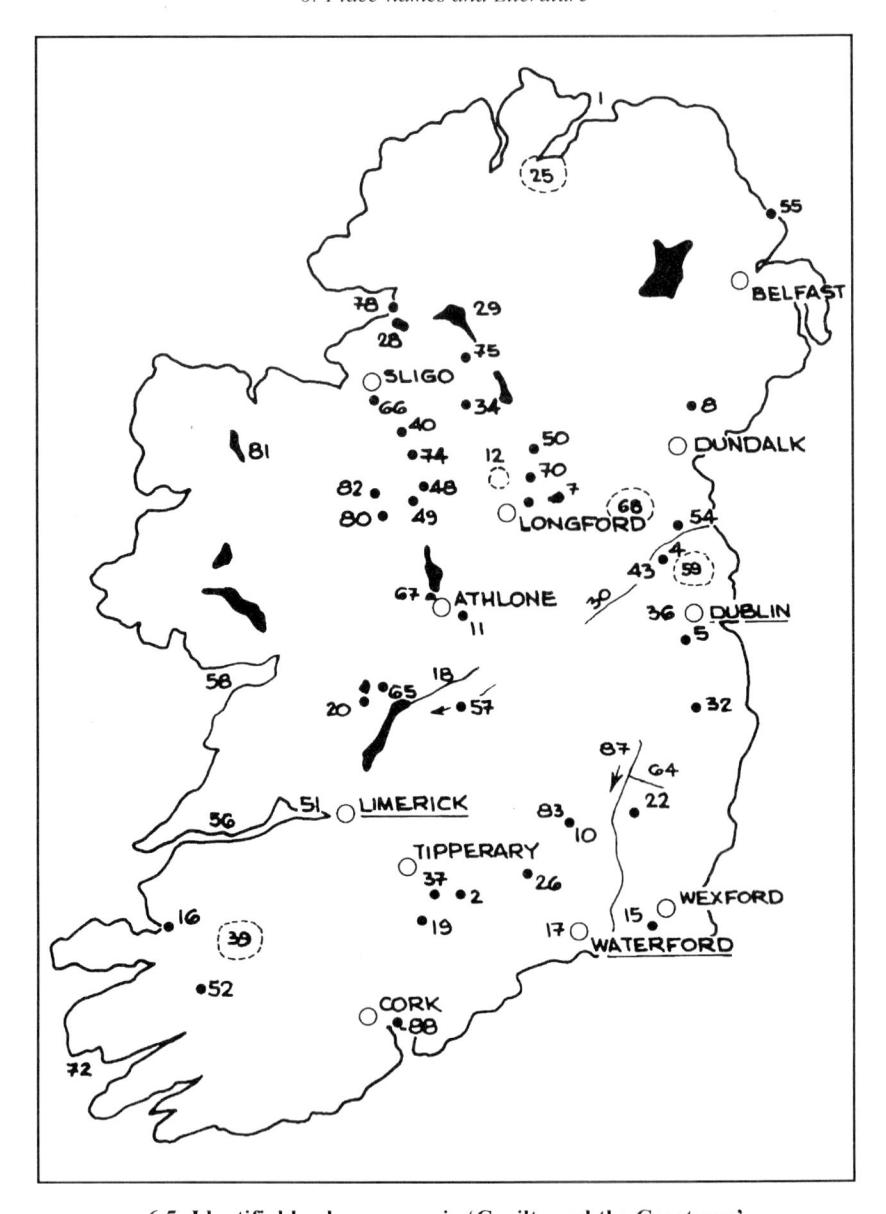

6.5: Identifiable place-names in 'Caoilte and the Creatures'
Identification is made in the first instance with the help of Hogan 1910, and the places plotted using current OS maps of Ireland. Unless underlined, names in capital letters are purely for reference purposes, and do not appear in the poem. Dotted circles approximately indicate the extent of named areas (Map: J. Renny)

6: Place-names and Literature

Key to places in 'Caoilte and the Creatures' identified on accompanying map (Fig. 6.5).

Place-names in **bold** type are those which we can identify with some certainty, and are shown on the map (except for 33 and 38). They are given in their form as found in the poem. Place-names in *italics* are those whose identification is uncertain, and which therefore do not appear on the map. English forms of place-names are in round brackets. Numbers in square brackets are line references. The numbering of the key follows the order of the place-names in the poem.

1 **Loch Foghail** (Loch Foyle) [8]
2 **Bodhamair**, ? near Cahir TPY [12]
3 *Siolar-ros* [40]; *Ros Iolarghlas* [40 sup.]
4 **Teamhair** (Tara) [2, 42, 44, 45, 46, 113, 156]
5 **Seisgeann Uairbheoil**, ? at Mount Seskin [80, 84]
6 *Fiodh Dhá Bheann* [119]
7 **Loch Seighleann** (Loch Sheelin) [120]
8 **Sliabh Cuilinn** (Slieve Gullion) [8]
9 *Buireann* [122]
10 **Fiodh Ghabhrain**, at Gowran KLK [123]
11 **Fiodh Fardhruim**, ? at Fardrum WMH [124]
12 **Coillte Craobh** (Kiltcreevagh townlands) [125]
13 *Druim <Dhá Bhraon>* [126]
14 *Iardomhan* [127]
15 **Carraig Dhobhair** (Carrickdover) WXF [128]
16 **Trá**i**gh Lí** (Tralee) [129]
17 **Port Láirge** (Waterford) [130]
18 **Brostnach Bhán** (? River Brosnach) [131]
19 **Carraig D[únáin]** (Carrigdownane) CRK [132]
20 **Eachtach** (Aughty Mountains) [133]
21 *Leitir Lonngharg* [134]; *Leitir Lomard* [134 sup.]
22 **Dún Aoibhthe** (Duneefy) CLW [136]
23 *Corraoibhthe* [136]
24 *Corann (Cladhach)* [137]
25 **Magh Foghail**, the area around Loch Foyle [138]
26 **Carraig na gClach**, on Sliabh na mBan [139]
27 *Fiodh Caondach* [140]
28 **Loch Mailbhe** (Loch Melvin) [141]
29 **Loch Eirne** (Loch Erne) [142]
30 *<Boinn>* (? River Boyne) [143]
31 *Dubhloch* [144]
32 *Magh Cualann* (? region in WCW) [145]
33 **Magh Fualainn** ? *recte* **Magh Tualainn**, in Éile, Lower Ormond [146]
34 **Gleann Aibhle** ? *recte* **Gleann Gaibhle** (Glen Gavlin) CVN [147]
35 *Seanaibhle* [148]
36 **Ath Cliath** (Dublin) [149]
37 **Crota Cliach** (Galtee Mountains) [150]
38 **Tráigh Dhá Bhan**, in or near Brega [151]
39 **Luachair Deagh[adh]**, region in KRY [152]; *Luachair <Adhair>* [152 sup.]; *Leitir <Uabhair>* [152 sup.]
40 **Ceas Chuirr** (Keshcorran) SGO [153]
41 *Leitir <Mhion>chuill* [154]
42 *Leitir Ruadh* [155]
(43 **Teamhair**, see 4 above.)
44 *Síth Dumha*, ? SGO [157]
45 *<Clach> Chorr* [158]
46 *Druim [an] Daimh* [159]
47 *Leathanmagh* [160]
48 **Léana Uair**, near River Uar at Elphir [161]
49 **Coill Ruadh**, otherwise **Coill Garbhruis**, near Tulsk [162]
50 **Cliabh Cleath** *recte* **Sliabh Cleath** (Slieve Glah), near Cavan [163]
51 **Luimneach** (Limerick) [164]
52 **Ath Lóich**, at Dunlow, near Killarney KRY [165]
53 *Móin Mhór* [166]
54 **Uamh Chnoghbha** (Cave of Knowth) MTH [167]
55 **Críoch Ollana** *recte* **Críoch Ollarbha**, round Larne Water [168]
56 **Sionna[nn]** (River Shannon) [169]
57 **Biorr** (River Birr) OFY [170]
58 **Cuan Ghailbhe** (Galway Harbour) [171]
59 **Magh Muirtheimhne**, area between River Boyne and Cuailgne Mountains in Brega [172]
60 *Fiodh Luaidhre* [173]
61 **Sídh Buidhe** (Shivey townland) TYE [174]
62 *Magh Moill* [175]
63 *Cnámhchoill* [176]
64 **Buireann** (River Burren) CLW [177]
65 *Dún Daighre* [178]
66 **Sliabh Dá Éan** (Slieve Daen), south of Sligo [179]
67 **Turlach Bruidhein** (Bree-oil), lochan near Athlone [180]
68 **Magh Balg** (Moybolgue), area in Kells MTH and CVN [181]
69 **Gránard** (Granard) LFD [182]
70 **Druing** ? (Dring), north of Granard [183]
71 *Mórchoill* [184]
72 **Dún na mBárc**, in Ballinskelligs Bay KRY [185]
73 *Gealtrácht* [186]
74 **Caladh Chairrge** ? (Rockingham), near Lough Key [187]
75 **Loch mac nÉan** (Lough Macnean) SGO and FER [189]
76 *Magh nOileán* [190]
77 *Meas an Chuill* [191]
78 **Eas Mhic Mhodhuirn** ? (Eas Ruaidh) or at Ballysadare SGO (map location favours former) [192]
79 *Gleann Smóil* [193]
80 **Ath Mogha** ? (Ballymoe) GLY [194]
81 **Loch Con** [195]
82 **Uamh Chruachan** (Cave of Rathcroghan) RSC [196]
83 **Sídh Ghabhláin Ghil** *recte* **Sídh Ghabhráin Ghil**, ? at Gowran KLK [197]; see 10 above.
84 *Druim Chaoin* [203]
85 *Gleann Dá Bhan* [215]
86 *Loch Lurgan* [216]
87 **Bearbh** (River Barrow) CLW, KLK and WFD [219]
88 **Inbhir Dhubhghlaise**, ? estuary of River Douglas at Cork [220]

Gaelic world by the capacious memories of narrators and the penmanship of many scribes. They were not tied to one particular location, even the 'original' location. As they travelled through the Gaelic world, the ballads took their topographical luggage with them.

The fact that the ballads belong to both Ireland and Scotland adds a special interest to this theme, since it seems plausible that certain ballads which were originally composed in Ireland were transmitted to Scotland and took root afresh in Scottish soil. The place-names which were integral to these ballads in their original form would have been relocated in Scotland, using as a bridge the core of generics shared within the Common Gaelic culture of both countries. Given the prevalent academic view that Ireland was the donor country, and Scotland the receptor, this is how one tends to interpret the *prima facie* evidence.

A fine example of probable east-to-west mobility is found in the case of *Laoidh Fhraoich* ('The Lay of Fraoch'). I have already argued that this ballad appears to have had its primary locus in Connacht, and that it was transmitted to Scotland at an early stage, as shown in its presence in the Book of the Dean of Lismore at the beginning of the sixteenth century. The theory of an Irish origin for the ballad does, however, sit somewhat uneasily with the absence of the text from almost all known collections of ballads in later Irish transmission. Again, I have suggested that, in the Irish context, the ballad (which narrated the death of Fraoch) was driven out, so to speak, by the alternative story of Fraoch's encounter with the monster (which claimed that, having been severely mauled, he was given a healing bath, and survived to go on further expeditions) (Meek 1984).

In Scotland, by contrast, the ballad was extremely popular, and relocated itself in several areas where there were *fraoch* place-names. In the eighteenth century it was strongly connected with Loch Freuchie (*Loch Fraochaidh*, either 'Loch of Heathery Place' or 'Stormy/Angry Loch'), near Amulree, in Perthshire. Near the south shore, the loch contains a small wooded island which was associated with the narrative, specifically as the location of the rowan tree which had healing powers, and whose berries Fraoch had to obtain for Meadhbh (Figures 6.6a and 6.6b).

Beyond Perthshire, the ballad took root at Loch Awe, where there was a convenient *fraoch eilean*, not far from the well-known mountain called Cruachan (Campbell 1872, 33). The latter could be equated with *Cruachú*, the name for Rathcroghan, the seat of Meadhbh and Ailill in Connacht. The poem was similarly contextualised in the Ross of Mull, where it was recited to J.F. Campbell in 1870 'on a heather knoll, near Ardfeenaig, almost within sight of Iona, Islay and Jura' (*ibid.*). Its domain in the Ross was described by

6: Place-names and Literature

6.6a: Loch Freuchie, showing the wooded island near the south shore of the loch

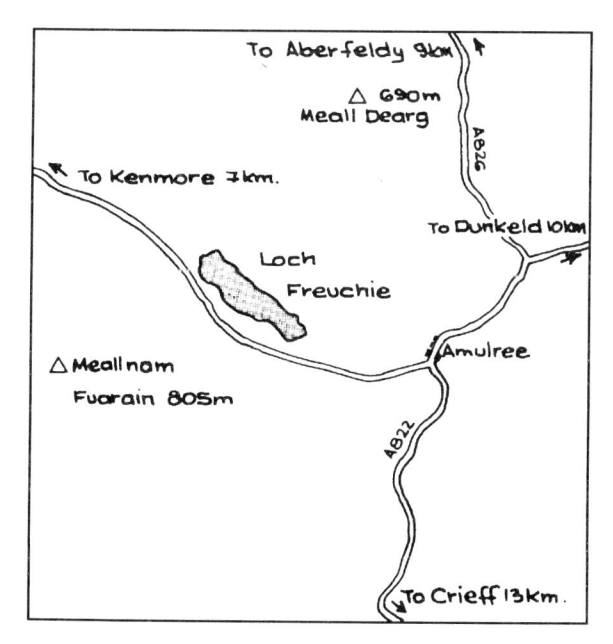

6.6b: The location of Loch Freuchie
(Map: J. Renny)

J. MacCormick and W. Muir in a booklet entitled *The Death of Fraoch*, published in 1887, as follows (Campbell and Henderson 1981, [xxvii–xxviii]):

'The local tradition is that Fraoch lived at *Suidhe*, near the village of Bunessan. Opposite him, in an oblique direction, lived Mev, through whose treachery Fraoch was slain; the place is still known as *larach tigh Meidhe* ("the site of Meidh's house"). The island where the rowan tree grew is called after her, *Eilean Mhain* (the isle of Main). It is right opposite to Bunessan.

'The local tradition asserts that the monster which guarded the rowan tree, and by which Fraoch was slain, was a great serpent; but we take leave to doubt this, because great serpents were not known in Scotland. We think that the creature was the *torc nimhe* (wild boar), which undoubtedly was common in the Highlands. Some who possess the tradition say that Fraoch was found dead with the heart of the beast in his hand, on the strand of the "Bay of the Heart" (*Camus a' Chridhe*). The bay is there to witness this, but we do not read in the poem that the monster's heart was really torn out.'

Within the ballads, then, we encounter a type of place-nomenclature which might be termed 'transferable toponymy' or 'portable place-names'! In this process, it is apparent that, when a narrative is transferred from one locality to another, it implants itself by grafting its key place-names on to its chosen location, sometimes selecting that location because appropriate place-names or, at the very least, appropriate place-name elements are already to be found there. The connections do not need to be specific or *literatim*; all that is required is a general similarity in one or more of the key onomastic elements, and a physical environment capable of sustaining the circumstances of the plot.

Thus, the version of *Laoidh Fhraoich* current in Perthshire and located at Loch Freuchie in the eighteenth century still retained *Loch Máigh* as the setting for action in the ballad, despite the fact that the location was now *Loch Fraochaidh*. A key verse in the text in the Book of the Dean of Lismore (q. 25) does, however, indicate that the ballad made provision for a parallel name for *Loch Máigh*, the parallel name apparently preserving the name of Fraoch (Meek 1982, 375):

> *Bhón bhás sin do fuair an fear –*
> *Loch Máigh, gé lean don loch –*
> *atá an t-ainm sin de go luan*
> *'ga ghairm anuas gus anos.*

(From that death that the hero endured – although Loch Máigh continued to be the name of the loch – it is called by that name until now and will be for ever.)

It is interesting that the quatrain does not specify the precise form of the *fraoch*-name for the loch; this would allow flexibility in selecting loch-sites for the ballad.

The existence of different names for a physical feature such as a loch may not have been too unusual; in the case of Loch Freuchie, the preamble to the version of *Laoidh Fhraoich* in the Gillies Collection indicates that 'The scene of the following poem is said to have been on the south shore, and on the Island near the south side of Loch-Cuaich, or Lochfraochy about two miles to the westward of Amalrie and eleven west from Dunkeld' (Gillies 1786, 107). In the late eighteenth century, therefore, Loch Freuchie was known to have two names, the one linking it with the glen lying to the northwest (*Gleann Cuaich*) and the other with *fraoch* (however understood). It is quite possible that *Loch Fraochaidh* was a secondary name.

The place-names *Cluain Fraoich* and *Carn Fraoich* were also preserved in the eighteenth-century Scottish text represented in Gillies. In this later context, however, they were apparently not regarded as 'official' place-names; they were descriptors of local significance – 'meadow of Fraoch' and 'cairn of Fraoch', making no demands other than to find a suitable field or mound relatively close at hand (*ibid.*, 108). In the case of *Gleann Síodh*, which, along with *Beann Ghulbainn*, was lugged around the Highlands and Islands in the ballad text, the need to find a specific onomastic location was rendered unnecessary by construing the name not as a place-name but as two common nouns: *gleann síthe*, meaning 'glen of peace' or 'fairy glen' (Meek 1990b, 345).

The evidence of Hebridean variants of *Laoidh Fhraoich* suggests that neither the maintenance nor the matching of place-name elements was deemed essential, at least within the wider context. In its utilisation of the place-name *Camus a' Chridhe*, which has no attestation within the ballad, the Mull version shows that a story could 'pull in' names and landmarks which originally had no connection with the place-names in the poem. It is also possible that a story could generate further place-names beyond those associated with the ballad text; thus place-names referring to the *torc* ('boar') are found in close proximity to the *Beann Ghulbainn* which is named in *Laoidh Dhiarmaid*. At the Glenshee location one finds *Loch an Tuirc* ('The Boar's Loch'), and in the Trossachs there is Brig o' Turk, which may (or may not) be part of a complex of place-names associated with the nearby *Beann Ghulbainn* (*ibid.*, 344–5). (See also Figure 6.4d.)

6: Place-names and Literature

The extent to which a ballad could supplement the existing place-names of a particular area is noteworthy. An incoming ballad could be accommodated in a new location even when there was no obvious 'matching' between its place-names and those of the adoptive area. To achieve this, the transmitters of the ballad would employ the rather cunning 'trick' of overlaying (or underlaying!) existing place-names with an alternative set of toponyms which were deemed essential to the narrative. We know that this happened in the case of *Laoidh Dhiarmaid*, when a *Beann Ghulbainn* had to be found or invented in an area where no such name had previously existed. For example, the story of Diarmaid took root in Skye, and in order to accommodate the story, it was popularly believed that *Beinn Tianabhaig* was once called *Beann Ghulbainn* (*ibid.*). A locality could thus come to have two sets of place-names, with both in use at the same time, the one set belonging to the physical geography of the area, and the other to the narratives which had come to be associated with it.

Given the nature of the ballad tradition, we need not suppose that this multi-faceted process of relocation was restricted to the transfer of material from Ireland to Scotland, or from one location to another within Scotland. In theory, poems could move from Scotland to Ireland, and doubtless did so, with the same implications for re-rooting and re-shaping of their place-name elements. Nor is it necessary to suppose that the 'toponymic transfer' was all one-way, with the incoming narrative acting as the main contributor to toponymic enrichment. It is at least possible that the opposite process could occur, and that an incoming narrative was sometimes adapted in such a way as to appear indigenous by equipping it with place-names which belonged to its adopted area.

That is why one must always be careful about making dogmatic claims about the countries or localities in which Gaelic ballads actually originated. The best we can do is to hazard an informed guess that the primary text of a ballad as we now know it has been shaped in a particular geographical environment. I like to think, and sometimes to argue (as I do here!), that *Laoidh Dhiarmaid* ('The Lay of Diarmaid') was composed in Glenshee, Perthshire. The *prima facie* evidence is compelling; like *Laoidh Fhraoich*, the ballad was hardly known in Ireland (the tale of Diarmaid's death being told in prose), and the version in the Book of the Dean names *Gleann Síodh* in its opening line – and I know of no suitable Glenshee in Ireland. But can I be absolutely sure that this proves a Scottish origin? Is it possible that a version of the ballad was received from Ireland, but that it was subjected to some skilful 'transplant surgery' to allow it to sound like a convincing Scottish product? Was it only the Scottish version, or even the Glenshee version, that was created in the shadow of that beautiful, snouty mountain? For the same reasons, can we be 100 percent sure that *Laoidh Fhraoich* was indeed composed in Ireland? Could it have been composed in Scotland, perhaps even in Perthshire, and then transmitted to Ireland, where a Roscommon version was produced? Did this New Irish Version become the dominant text, later returning to Scotland replete with Roscommon placenames? Or was the ballad composed in Scotland by a Scotsman who had an intimate knowledge of this district of Roscommon? We cannot find the answers to these questions, at least in the present state of our knowledge.

Conclusion: place-names or symbols?

Given such uncertainties, we must exercise due caution in interpreting place-names and their significance within a literary context. We even have to face up to one of the most perplexing questions which one can ask in this field: 'When is a place-name not a place-name?' The answer (for present purposes) is 'When it is a symbol', or, indeed, 'When it is an icon', representing a point or event in time or place, within the mental and narratological 'map' of the composer. Even place-names which look, and at one level are, 'real' enough can be uprooted and invested with a significance far beyond the merely topographical. If place-names are indeed movable and adaptable, and if they can be invested with a significance beyond the requirements of the earthbound traveller, we cannot be too careful in analysing the sign-posts. If we do not read them correctly, there is a chance that we will never arrive at our destinations, or even at our starting-points!

Acknowledgements

I am very grateful to my colleague, Dr James Grant, for alerting me to the existence of Beinn Ghuilbin to the north of Aviemore, and for providing a photograph of the hill (Fig. 6.4a); and also to Mrs Sheila McGregor, Edinburgh, for providing further references to Scottish hills of this name.

Bibliography

Almqvist, Bo, Ó Catháin, Séamas and Ó Héalaí, Pádraig (eds.), *The Heroic Process: Form, Function and Fantasy in Folk Epic* (Dublin, 1987).
Bruford, A., *Gaelic Folk-tales and Mediaeval Romances* (Dublin, 1969).
Campbell, J.F. (ed.), *Leabhar na Féinne* (London, 1872).
Campbell, J.F. and Henderson, George, *The Celtic Dragon Myth* (North Hollywood, California, 1981).
Christiansen, R. Th., *The Vikings and the Viking Wars in Irish and Gaelic Tradition* (Oslo, 1931).

Gillies, J. (ed.), *Sean Dain agus Orain Ghaidhealach: A Collection of Ancient and Modern Gaelic Poems and Songs* (Perth, 1786).

Gillies, W., 'Arthur in Gaelic Tradition: Part 1: Folktales and Ballads', *Cambridge Medieval Celtic Studies*, 2 (1981), 47–72.

Hogan, E., *Onomasticon Goedelicum* (Dublin, 1910, reprinted 1993).

MacInnes, J., 'Twentieth Century Recordings of Scottish Gaelic Heroic Ballads', in Almqvist *et al.* (1987), 101–30.

McCone, K.R., 'Werewolves, Cyclops, Díberga, and Fíanna: Juvenile Delinquency in Early Ireland', *Cambridge Medieval Celtic Studies*, 12 (1986), 1–22.

Meek, D.E., 'The Corpus of Heroic Verse in the Book of the Dean of Lismore' (unpublished Ph.D. thesis, University of Glasgow, 1982).

Meek, D.E., 'Táin Bó Fraích and Other 'Fráech' Texts: A Study in Thematic Relationships, Part I', *Cambridge Medieval Celtic Studies*, 7 (1984), 1–37.

Meek, D.E., 'The Banners of the *Fian* in Gaelic Ballad Tradition', *Cambridge Medieval Celtic Studies*, 11 (1986), 29–69.

Meek, D.E., 'Development and Degeneration in Gaelic Ballad Texts', in Almqvist *et al.* (1987), 131–60.

Meek, D.E., 'The Scots–Gaelic Scribes of Late Medieval Perthshire', in *Bryght Lanternis*, eds. M. Spiller and D. McClure (Aberdeen, 1990a), 387–404.

Meek, D.E., 'The Death of Diarmaid in Scottish and Irish Tradition', *Celtica*, 20 (1990b), 335–61.

Meek, D.E., 'The Gaelic Ballads of Scotland: Creativity and Adaptation', in *Ossian Revisited*, ed. H. Gaskill (Edinburgh, 1991).

Murphy, G., *The Ossianic Lore and Romantic Tales of Medieval Ireland* (Dublin, 1961).

Ó hÓgáin, D., *Fionn mac Cumhaill: Images of the Gaelic Hero* (Dublin, 1988).

7

Gwaun Henllan – the Oldest Recorded Meadow in Wales?

Heather James

In 1993 Celtic Energy submitted a planning application to the then Dyfed County Council to carry out open cast coaling on an area termed by them Tir Dafydd – an adaptation of a tenement name within the area (see Figs. 7.1 and 7.3). This area of south-east Carmarthenshire has had a long experience of open cast coal mining of the rich anthracite, a technique that has now almost wholly replaced deep mining. But the application was refused by the County Council; among the reasons cited for the decision was:

> The proposed development if approved would physically remove a significant area of land which, as a result of the investigations carried out as part of the planning application, has been identified as a landscape of high value or very high value ecologically, archaeologically and historically (Dyfed County Council C6/295).

Celtic Energy appealed to the Secretary of State for Wales against the decision and a Public Inquiry was held into their appeal early in 1996. There were three parties in the Inquiry: Celtic Energy, Dyfed County Council (as the Mineral Planning Authority) and, collectively, the communities of Llandybïe, Llanfihangel Aberbythych and Gorslas. In November 1996, the Secretary of State announced that he had accepted the Recommendations of his Inspector, Mr David Shears, and the appeal was dismissed. Dyfed Archaeological Trust (DAT) was involved in the preparation of the original Environmental Statement for British Coal Open Cast and was also retained as expert witness by the mineral planning authority – Dyfed County Council. All the documentation relating to the sites and planning history is deposited in the County *Sites and Monuments Record,* maintained by DAT (prn 32118 & 32119).

Although taking their place amongst many other issues, the importance of

7: Gwaun Henllan – *the Oldest Recorded Meadow in Wales?*

7.1: Location map of Llandybïe, Carmarthenshire

place-names as part of the cultural heritage of an area was argued in the Inquiry and upheld by the Inspector. One name and location in particular was central to the case – that of two contiguous fields of rough pasture, rich botanically, both called *Gwaun Henllan*. They lie within the middle of the proposed coaling area. Dyfed County Council argued that these fields were in the same use and location as the ninth century *guoun hen llan*, named as one of the bounds of an estate called *mainaur med diminih*, in Chad 6, a marginal note in early Welsh, one of a group recording land grants to the monastery of Teilo, to be found on the pages of an eighth-century Gospel Book, the so-called Lichfield Gospels, or Book of St Chad.

The precision of this identification, and its significance as a reason for preserving the fields themselves was disputed by an archaeologist retained by Celtic Energy – A.G. Marvell, of the Contracts Section of Glamorgan-Gwent Archaeological Trust. The author of this chapter was retained by Dyfed County Council as expert witness, and the proofs of evidence, rebuttals, counter arguments and cross-examination all prompted the speedy execution of a great deal of primary research to prove the case which might otherwise have not been thought necessary to carry out in less adversarial circumstances.

A definitive study of the Welsh marginalia has been published by Jenkins and Owen, 1983 and 1984. They conclude that the now incomplete Gospel Book, of eighth-century origin, was at St Teilo's monastery at Llandeilo Fawr, in the Tywi valley, Carmarthenshire, by the beginning of the ninth century. It was removed from there (by Mercian raids?) and has been at Lichfield since the mid-tenth century at the earliest, certainly the eleventh century: hence its title of The Lichfield Gospels or the Book of St Chad, and Melville Richards's re-naming as the Llandeilo Gospels, or Book of Teilo (Richards 1973). There are six marginal entries (Chad 1–6), the first in Latin, the remainder in Welsh, dating from the early ninth to perhaps as late as the early tenth centuries. Chad 2, the so-called 'Surexit' memorandum, is the earliest written Welsh text. Chad 6 differs from the Celtic charter form of 3 and 4 (for which see Davies 1982) in that it records the *nobilitas*, (Welsh *braint* ?) and *mensura* (extent) of *mainaur med diminih*. This was identified as *Maenor Meddynfych*, an estate centring on a farm close to modern Ammanford in Llandybïe parish, Carmarthenshire, NGR SN629132 (for a note on the development of the place-name, see Williams 1933–35). A full study has been made of this territorial unit, and other Chad marginalia by the late Glanville Jones (Jones 1972). He forwarded a developed version of his 1972 map of *Maenor Meddynfych* to me in 1988, in correspondence about the area, and this is reproduced in Figure 7.2.

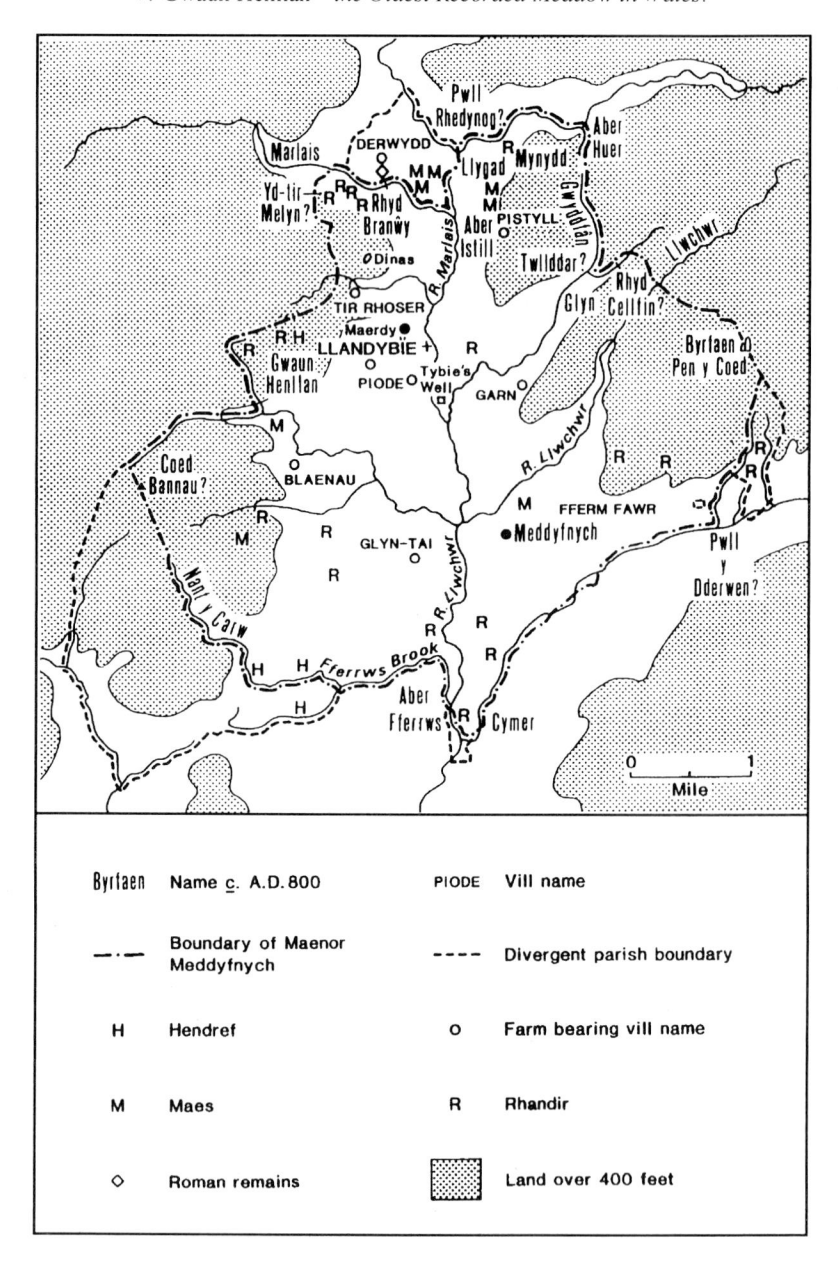

7: Gwaun Henllan – the Oldest Recorded Meadow in Wales?

Byrfaen	Name c. A.D. 800	PIODE	Vill name
—·—	Boundary of Maenor Meddyfnych	- - - -	Divergent parish boundary
H	Hendref	O	Farm bearing vill name
M	Maes	R	Rhandir
◇	Roman remains	▓	Land over 400 feet

7.2: The medieval parish of Llandybïe, with the bounds of *Maenor Meddynfych*
(Copyright the estate of Professor G.R. Jones)

7: Gwaun Henllan – the Oldest Recorded Meadow in Wales?

Many of the place-names given for the bounds of the *maenor* can be identified with stream names, or other features on the boundary of the ecclesiastical parish of Llandybïe. The bounds of the *maenor* do not exactly coincide with the medieval parish. Derwydd to the north-west, and an area of upland to the south-west were included in the parish. The former was probably an early high-status site with a continuing importance throughout the Middle Ages and beyond; the latter area may represent common upland which was later apportioned between adjacent parishes. Parish boundaries were not finally fixed in Wales until the thirteenth century (Williams 1976, 15) and many parochial units, particularly in the Welshries of Anglo–Norman lords or in areas still under the control of the native Welsh princes, were simply a continuation of earlier administrative or territorial boundaries. Glanville Jones saw the area as a multiple estate, with Meddynfych as the *llys*, or court, as its centre. He noted numerous relict traces of former sharelands, both of arable and meadow from the Llandybïe Tithe Map and its accompanying Schedule of 1839. He singled out *Gwaun henllan* as a possible indicator of the antiquity of communal agrarian arrangements even by the ninth century:

> A productive upland meadow, this lay at an altitude of about 500 feet, a mile or so west of the lowland settlement which clearly was already known by the name Henllan (Old Church). That a distant upland meadow should be thus attributed to the Old Church in itself hints that there was a hamlet continuity here even by this date (Jones 1988).

Local historians have preferred to see the *henllan* element of the place name as indicating a former chapel site close to the meadow, within the present day Glyn-yr-Henllan Farm (Roberts 1939, 30). Evidence presented at the Inquiry revealed a rich folklore for the whole area presented through its toponymy.

Celtic Energy's archaeologist challenged the precise equation of the Tithe Map and today's *Gwaun henllan* with the ninth-century meadow. *Waun* and *henllan* names, he asserted, could be found in other areas within the parish and outside; the description *henllan* covered the whole north bank of the Afon Lash in the parishes of Llanfihangel Aberbythych as well as Llandybïe. For a period in the eighteenth century and the early nineteenth century, Glyn Farm in Llanfihangel Aberybythych was known as Glyn Henllan. There is of course no particular significance to the widespread prevalence of the *waun* or *gwaun* element; it is a common form with a fairly wide range of meanings. GPC, which notes *guoun hen llan* as the earliest form, gives for both *gwaun* and the plural *gweunydd* 'high and wet level ground, moorland, heath, low-lying marshy ground, meadow'.

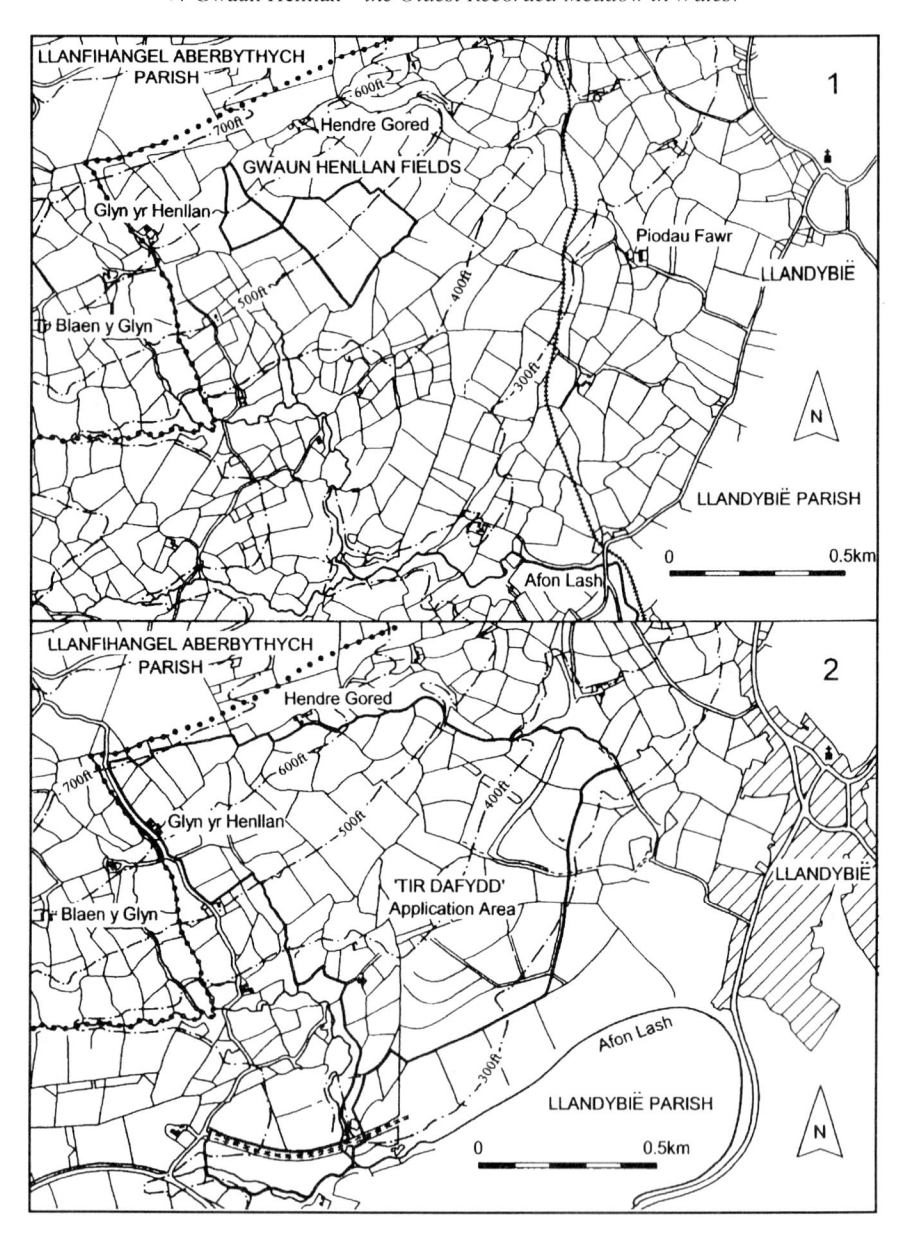

7: Gwaun Henllan – *the Oldest Recorded Meadow in Wales?*

7.3: The parish of Llandybïe:
Map 1: Field patterns pre-open cast coaling west of Llandybïe;
Map 2: Reclaimed open cast land and active coaling areas

7: Gwaun Henllan – the Oldest Recorded Meadow in Wales?

This claim was rebutted by means of a count and analysis of all the field-names and field-name elements in the Tithe Schedule for Llandybïe parish. Some 80 percent of the 4,086 numbered field entries have a name, so the sample of named fields is a large one. The commonest element is *cae*, which means 'field', next come the *waun* place-names. There are 1,129 field-names with the *cae* element and 415 with *waun* or *gwaun*. It is possible to separate this large *gwaun* sample into two groups – those meadows, often small to medium in size, which form a part of the farms which have come into existence from the sixteenth century onwards, often only called *waun*, and those which seem likely to be of earlier origin. The latter are characterised by having individual names, most of which can be traced back to sixteenth- and seventeenth-century documents, names which occur across more than one Tithe Map field, and where there is often a trace still in 1839 of shared meadowland. They are also larger in area. *Henllan* as a name or as an element was only found to occur twice in the whole of the Llandybïe Tithe Schedule, in fields 1860 on Glyn-yr-Henllan Farm and 1880 on Hendre Gored Farm, where it is linked to a *gwaun* name – *Gwaun Henllan*.

Nor can *henllan* be seen as a common name or element in Carmarthenshire, as was also asserted by Celtic Energy. Here use was made of Carmarthenshire Antiquarian Society's computerised Place Names Index (James 1990). It was surprising in fact to see how uncommon *henllan* is, and this has caused me to reassess my earlier, cautious, interpretation of the *Gwaun Henllan* name as perhaps meaning simply 'meadow of the old enclosure' rather than 'meadow of the old church' – although it is still uncertain whether Llandybïe church itself was 'the old church' as Glanville Jones thought, or, as local historians believe, a lost chapel close to Glyn-yr-Henllan Farm, to which *Gwaun henllan* would have been more closely attached.

Most of the parish of Llandybïe and the adjoining parish of Llanfihangel Aberbythych was owned by the two leading landowning families of seventeenth- and eighteenth-century Carmarthenshire – the Vaughans of Golden Grove and the Rices of Dynevor, with the fortunate result that the Tir Dafydd area is well documented from the sixteenth century onwards by deeds, rentals and maps. Both Glyn-yr-Henllan and Hendre Gored Farms therefore are depicted in late eighteenth-century estate map books. These maps of the 1780s were used at the Inquiry to show how little the field pattern had changed since that date. A 1783 map of Hendre Gored Farm, which by that date contained the eastern part of the large *gwaun henllan* meadow, is of particular interest in that three unfenced strips are shown across the north of the field. One is labelled 'Castell-y-graig – the hay'; south of that 'Wm. Morris of Garn – the hay only' and the strip below that unnamed. Castell-y-graig and the Garn were neighbouring farms and here we have the last, relic, traces of an earlier shared use of the meadow. (CRO CV *Golden Grove Map Book* 84–5.)

Beyond the second half of the eighteenth century *Gwaun Henllan* had to be traced through deeds and rentals. Both the Cawdor/Vaughan collections at Carmarthen Record Office and the Dynevor collection at the National Library of Wales contain sixteenth- and seventeenth-century deeds which show how many of their individual farms, leased out to tenants, were built up by exchanging and consolidating strips of land from former sharelands. Such was *Gwaun Henllan*. A deed of 1547–1553 records the release and quitclaim of parcels of land in a place called *Goweyn Henllan* and names the owner of adjacent lands to the south (CRO CV 124/8630). A grant of 1557 refers to half a messuage at 'Rosser' and part of a meadow at Henllan (CRO CV 54/6167). Tir Rosser, an individual farm by the seventeenth century at least, perhaps a hamlet in origin, gave its name to one of the component townships of the parish. It lies to the north-east of *Gwaun Henllan*. Glanville Jones glossed Tir Rosser as *Tir yr Orsedd* – 'land of the mound or court or throne' (W. *gorsedd*). This grant describes the meadow land as lying between the lands of Thomas Howell, William Philip, Rees Morgan John, David ap Ieuan Gwillim Boole and the lands of the crown. This latter piece of land must be part of the former demesne lands of Sir Rhys ap Gruffydd, escheated to the crown when he was accused of treason and executed in 1531. Not until the second half of the sixteenth century did the Rice family, as they became known, recover their lands and fortune.

There are no individual references between the ninth century and the sixteenth, but it is only in the Duchy of Lancaster records where any reference might be expected, such is the paucity of surviving medieval Welsh rentals, court rolls, inquisition surveys etc. Llandybïe parish lay within the commote of Iscennen, where native Welsh law, kinship, and tenurial structures remained relatively unaffected by Anglo–Norman conquest. It became part of the Lordship of Kidwelly and thus the Duchy of Lancaster lands only in the fourteenth century. There can be little doubt therefore that the ninth-century *guon hen llan* is at the same location as today's rough grazing land on Glyn yr Henllan and Hendre Gored farms.

When these two farms came into existence as individual holdings is less easy to say, but there are traces of older arable sharelands within the former farm at least. Just east of today's Glyn-yr-Henllan farm, north of *Gwaun Henllan*, is a long, narrow field named in the Tithe Schedule as *rhandir*, literally 'shareland' but translated in a seventeenth-century Welsh dictionary by Thomas Jones as 'a Lot or Hereditary Part' (Jones T. 1977). Glanville Jones plotted this place-name element from the Llandybïe Tithe Schedule (see Fig. 7.2), taking them as relic traces of former *rhandiroedd* or arable sharelands. Such a shareland, in the idealised constructs of the Welsh Lawbooks, would have contained some 300 acres of Welsh measure for arable, pasture and fuel, belonging to a *gwely* or kinship group.

In his report conclusions and recommendations, the Inspector, Mr David Shears, accepted the arguments made for the antiquity and value of the historic landscape elements of the Tir Dafydd area. His conclusions on the importance of the area's toponymy as an integral part of its cultural heritage are of particular interest in the context of this volume:

> 9.30. The site would appear to meet several of these [Criteria for inclusion in the Cadw/CCW *Register of Parks, Gardens and Historic Landscapes*] in that it is a relict landscape, little changed over a considerable period of time, and that it shows historic continuity (time depth) and a degree of integrity and coherence. For these reasons I am satisfied that considerable weight should be placed on the need for the preservation of the Gwaun Henllan landscape. Merely to record place-names that used to exist can serve little purpose without the preservation *in situ* of the landscape to which they refer, especially as in this case that name records a historic landuse that can be directly linked to the present day.

and again:

> 9.31. Indicative of the importance of this landscape for cultural as well as historic reasons is the fact that all signs of numerous Welsh place names and localities have already been obliterated from the landscape in the general area through previous opencasting activities. This is evidently of considerable concern to the local communities, who are still predominantly Welsh speaking. This context highlights the need to preserve what remains, especially as it can be linked to the earliest origins of the Welsh language...it is clear that the proposed development would result in the total loss for all time of a landscape of considerable historic and cultural significance.
> (Annexe to Welsh Office Planning Division letter PP121-98-001, 7 November 1996)

The loss from previous open casting is well illustrated by the previously opencasted and reclaimed area east of the Tir Dafydd area. This once contained a fine seventeenth-century gentry house called Piode fawr (see Fig. 7.3, Map 1). Alerted, in the early 1970s, by the onset of opencast coaling in this area, formerly occupied by a deep mine – Llandybïe colliery – the then County Archivist, Major Francis Jones and the former Secretary of the Welsh Royal Commission, Peter Smith (gathering material for his monumental *Houses of the Welsh Countryside*) visited and discovered 'this

7: Gwaun Henllan – the Oldest Recorded Meadow in Wales?

little gem of an early Renaissance house' (Smith, 1974, 69). The house was boarded up and left within the opencast site, but theft and vandalism took its toll and it was demolished. No trace of its former existence is reflected in the very new, open, empty landscape which was created over the reclaimed site.

Here then was not just the loss of a fine house which could and should have been saved but also the probable buried archaeological traces of its medieval predecessors. For Piode gave its name to one of the eight 'hamlets' into which the parish was divided in the late seventeenth century, as Edward Llwyd records. He also described Piode as 'an ancient seat belonging to Arthur Gwyn. Esqr.' (Fisher 1917). In many other areas of Carmarthenshire besides the parish of Llandybïe, where medieval Welsh settlement patterns were little affected by Anglo–Norman reorganisation, minor gentry seats may well occupy medieval sites which were the centre of the hereditary lands of a *gwely* or extended kinship group.

We have seen above (p. 173) how the hamlet, or township of Derwydd was added to *Maenor Meddynfych* to be encompassed within the parish of Llandybie. Remove it and we are left with seven townships. Here, as in so many other ways, Glanville Jones led the way. The south-Walian *Llyfr Blegwryd* redaction of the Welsh Laws (for its close connections with the Tywi Valley see Jenkins 1986, xxii) states that there are seven townships (*trefi*) to a lowland *maenor*. Glanville Jones noted how seven townships survived as administrative units in Llandyssul, north Carmarthenshire, as well as Llandybïe and there are other examples (Jones 1972). In the late thirteenth/early fourteenth century then this legal construct had a territorial and administrative reality on the ground. The rare beam of light provided by the eighth- and ninth-century records of estates in Carmarthenshire, written on the leaves of the Llandeilo Gospels, suggest a continuation and development of ancient, well-established tenurial and territorial arrangements. The reasons for their survival may well lie in their incorporation and development in political and administrative structures of the native Welsh polity of the land of Deheubarth under the Lord Rhys in the twelfth century. These structures were retained by the English crown following the Edwardian conquest in order to counter the power of the Marcher Lords and strengthen the crown's claim to be the inheritor in title of the powers of the Welsh princes (Davies, R.R. 1978).

In the absence of early documentary sources, the surviving landscape and its toponymy represents one of our principal sources of information. I am currently pursuing the landscape history of the component Llandybïe townships/hamlets/kin or clan lands as we may variously term them at different periods of time. The case of *Gwaun-henllan* shows how important the place name evidence can be, particularly if analysed within its component landscape. The complete loss of large blocs of that landscape through open cast should, I believe, be strongly contested in every case – but only rarely, unfortunately, will a ninth-century meadow named in the earliest written Welsh provide the heart of the conservation case!

Abbreviations

CRO   Carmarthenshire Record Office.
GPC   *Geiriadur Prifysgol Cymru: A Dictionary of the Welsh Language* Caerdydd Gwasg Prifysgol Cymru.
DAT SMR   Sites and Monuments Record of Dyfed Archaeological Trust; prn = primary record number.

Bibliography

Davies, R.R., *Lordship and Society in the March of Wales* (1978).
Davies, W., 'The Latin Charter-Tradition in Western Britain, Brittany and Ireland in the Early Medieval Period', in *Ireland in Early Medieval Europe: Studies in Memory of Kathleen Hughes*, ed. D. Whitelock (1982), 258–80.
Fisher, J. (ed.), 'Lhuydiana, including an account of the parish of Llandybie', Appendix 4 to *Tours in Wales* (1804–1813), Richard Fenton (Cambrian Archaeological Association, supplemental volume for 1917), 336–7.
James, T., 'The Carmarthenshire Place-Name Survey', *The Carmarthenshire Antiquary*, 26 (1990), 91–4.
Jenkins, D. and Owen, M.E., 'The Welsh Marginalia in the Lichfield Gospels, Parts I and Part II: The "Surexit" Memorandum', *Cambridge Medieval Celtic Studies*, 5 (1983, 1984), 37–66 and 7, 91–120.
Jones, G.R.J., 'Post-Roman Wales' in *The Agrarian History of England and Wales*, I, ed. H.P.R. Finberg (1972), Part 2, 308–11 and Fig. 43.
Jones, T., *The Welsh–English Dictionary of 1688* (facsimile reprint by The Black Pig Press, Llanwrda, 1977).
Richards, M., 'The "Lichfield" Gospels (Book of "Saint Chad")', *National Library of Wales Journal*, 18 (1973), 135–42.
Roberts, G., *Hanes Plwyf Llandybïe* (1939).
Smith, P., 'Historical Domestic Architecture in Dyfed: An Outline', in *Carmarthenshire Studies: Essays presented to Major Francis Jones*, eds. T. Barnes and N. Yates (Carmarthenshire County Council, 1974).
Williams, G., *The Welsh Church from Conquest to Reformation* (Cardiff, 1976).
Williams, I., 'Meddyfnych', *Bulletin Board of Celtic Studies*, VII (1933–35), 369–70.

Afterword

Professor W.F.H. Nicolaisen

The Conference, the *acta* of which are reflected in this volume, may not have been the very first to involve scholars working in different disciplines in the discussion of name studies. International onomastic congresses often bring together people whose perspectives on the utilisation of names in their respective areas of research tend to vary widely, but the St Andrews meeting on 17 February, 1996, was obviously more severely focused in this respect than the more diffuse international gatherings normally are, and was undoubtedly within a Scottish context the first symposium to concentrate solely on some of the multiple uses that scholars can profitably make of names. By restricting the central theme further to the names of places, the several presentations were orchestrated to display a kind of pleasant cohesion not often encountered in meetings concerned with name studies, thus offering a fascinating diversity within its fundamental unity.

Since the chapters in this volume speak for themselves, there is no need to summarise their substance and approaches. Each of them confirms the rich source toponymic evidence can be if exploited with knowledge and care, whether one investigates it in conjunction with a visual inspection of actual features in the landscape as part of a quest for the etymologies of the elements place-names contain; explores the ways in which place-names present the (Scottish) historian with pointers, pitfalls and puzzles; surveys the relevant Scottish place-nomenclature with the aim of discovering what light it can throw on the historical development of certain aspects of Scottish Gaelic; examines the function of place-names in some genres of traditional Scottish Gaelic verse; utilises them as pointers to important ancient archaeological sites; or combines archaeological and onomastic strategies in a joint venture designed to explore the place-names of a limited area more fully than separate, unconnected investigations would have been able to do. Naturally, the approaches offered in these chapters cannot be expected to be comprehensive, but they nevertheless afford a glimpse of what can be done once the great variety of tactics has been recognised which can unlock the abundant resources which place-names have to offer, either by themselves or

Afterword

complementary to other, non-onomastic evidence. In the process, names are freed from a purely linguistic approach while being explored systematically, rigorously and imaginatively in the development of their truly inter-disciplinary potential.

It was this strong hint of team-work and co-operation in the future utilisation of Scottish place-names which made this conference so memorable and which is therefore echoed in this volume. For this reason, it was almost inevitable that the meeting culminated in the founding of a Scottish Place-Name Society in which the various strands of place-name studies in Scotland will be happily represented by its members. Modern, technical advances have made much more sophisticated approaches to the gathering, storing, analysis, and publication of place-name material possible, and the essentially inter-disciplinary character of that material will be the basis not only for inter-departmental co-operation within single academic institutions, but also for joint undertakings among more than one Scottish university. In such co-ordinated efforts the study of place-names is therefore likely to get much closer to realising its full potential which has lain dormant for so long, and to take its rightful place at the cutting edge of modern academic research. The large audience which attended the St Andrews Conference certainly thought so, and it is hoped that the readers of this volume will feel the same.

Index

The following index is simply a rough guide to the place-names and place-name elements discussed in the book, and is no way comprehensive.

aber, 55, 56, 57, 59
Abercarf, 56
Aberdeen, 31, 35, 50, 53, 54, 75
Aberdeenshire, 20, 31, 35, 36, 49, 51, 53
Aberlady, 56
Aberlemno, 39
Aberlessic, 56
Aberlosk, 56
Abermelc, 56
abhainn, 48
Abriachan, 62
Achnagairn, 26
Achnagullan, 27
Achpopuli, 62
Afon Lash, 173
Aikey Fair, 51
Airthrie, 31
Aith, 139
Alde, 31
Aldreth, 95, 100
Allen, 159
Almhain, 159
Alt Gellagach, 27
Altrie, 31
Ammanford, 171
Amulree, 162
Annandale, 56, 72
Aquhorthies, 5
Ardencaple, 26, 50
Ardnamoghill, 25
Ardnapreaghaun, 25, 26
Arran, 17, 47
Assaroe, 158, 159
Assendon, 85, 89, 99
Athelney, 92, 93, 100
Auchencairn, 23, 48
Auchgower, 50
Auchnabo, 50

Auchnabony, 30
Auchnagorth, 50
Auchterhouse Hill, 67
Aughnagon, 26
Aviemore, 154, 157
Ayrshire, 59, 68, 70, 72

Babbet, 68
Bada na Breasach, 27
Badenoch, 38
baile, 28, 56, 73
Baile na Beirbhe, 159
Baile na gCléireach, 49
Balbriggan, 33
Balerno, 38, 39, 42
Balernock, 38
Ballinabranagh, 25, 26
Ballinagore, 26
Ballincrea, 23
Ballindean, 21, 48
Ballingry, 8
Ballinkilleen, 33
Ballyandreen, 33
Ballybranigan, 22
Ballynacleragh, 25, 49
Ballynagaul, 26
Balmanno, 40
Balmerino, 21, 38, 39
Balnagore, 27
Baltasound, 123, 140, 141
Balvaird, 23
Balveny, 23
Banchory, 5
bangor, 113
Barnamon, 30, 50
Barnweill, 68
Barrhead, 70
Bayvil, 113

182

bealach, 4
Beann Ghulbainn, 7, 153, 154, 156, 158, 165, 166
Beauly, 8
bedd, 113
Begbie, 70
beinn, 56
Beinn Ghuilbin, 154, 157
Beinn Tianabhaig, 166
Belhelvie, 22
Bellendean, 21
Belmont, 129, 140, 142
Ben Gulapin, 154
Ben Gullipen, 7, 154
Ben Vane, 23
Benaveoch, 29, 50
Benbulben, 154, 156
Benderloch, 26, 28
Bengairn, 29
beorg, 78, 80, 81, 98
berg, 78, 80, 81, 98
Bergen, 159
Berkshire, 77, 92
berry, 106
-*bie*, -*by*, 72
Biffie, 31, 33, 36
Billington, 78, 79, 98
Binny, 31, 36
Birkhill, 72
Birkland Barrow, 78, 98
Birsay, 135
Blackford, 62
blaen, 73
Blairnagobber, 30
Blairnavaid, 27
Blantyre, 8
Bleadney, 94, 95, 100
Blegbie, 70
Blegbie Hill, 65
Bochastle, 26
bòg, 4
Bolney, 94
bólstaðr, 141
Bonskeid, 26
Boolanave, 25
Borenich, 26, 28
Bornish, 122
boþl, 67, 68, 69

botl, 67
Bourtie, 31
Braiklay, 52
Brig o' Turk, 165
Bromsberrow, 78, 80, 98
Brookpoint, 126, 128, 129
Brough of Birsay, 134
Brownber, 78, 98
Brunt Hill, 65
Buckinghamshire, 76
Buildwas, 93, 94, 100
Bunchrew, 26
Bunessan, 164
burg, 67, 69
burh, 67, 69
Busbie, 72

c(h)astel, 73
Cabrach, The, 39
cadair, 106
cae, 103, 116, 117, 118, 175
Cae'r dderwen, 103
caer, 6, 73, 102, 103, 106, 108, 113
Caerau, 113
Caerleon, 115
Caer-lleon, 115
Cairnagad, 27
Cairnbrogie, 31, 36
Cambo, 38, 42
Cammo, 38
Campsie Fells, 72
Camus a' Chridhe ('Bay of the Heart'), 164
carden, 55
Cardiff, 103
Cardigan, 103
Cardiganshire, 102
Carew, 103, 104
Carmarthen, 6, 103, 106, 112, 115, 176
Carmarthen Priory, 115
Carmarthenshire, 102, 108, 115, 116, 169, 170, 171, 175, 178
càrn, 6, 56
Carn Fraoich, 150, 151, 154, 158, 165
Carnfree, 151
Carrignavar, 25
Carstairs, 73
castell, 103, 105, 106, 108, 109, 111

Index

Castell Draenog, 6, 108, 110
Castell-y-graig, 175
castle, 103, 105, 108, 109, 111
Castle Combe, 99
Castle Enigan, 22
Castlemilk, 56
Castleplunket, 151
Chelmorton, 78
chester, 103
Chiltern Hills, 77, 78, 85, 88
Chinnor, 85, 99
cill-, 27
Cilsant, 115
Clarabad, 70
Clarembald, 70
Clermiston, 70
Clibberswick, 132
Clonegall, 26
Cloneygowan, 26
Cloondanagh, 20
Cluain Fraoich, 150, 151, 165
Cluny, 38
Clyde, 59
Clydesdale, 56, 68, 72
Co. Roscommon, 151
Coigach, 39
coirthe, 5
Colvedale, 130, 141
Combor, 19
Connacht, 17, 18, 148, 149, 150, 151, 152, 162
Coolcullen, 23
Corbie, 72
Corsbie, 70
Corstorphin, 8
Cothi, 117
Cowpe, 91, 99
Craigie, 31, 36
Craigowl, 67
Crail, 3
Crawford, 72
Creechbarrow, 81, 82, 98
Crookbarrow, 81, 82, 98
Crosbie, 72
Crouch Hill, 81, 82, 98
Cruachan, 162
Cruachú, 162
crūc, 81, 82, 98

crūg, 98
Crutch, 81, 82, 98
Cuddesdon, 78, 79, 98
cumb, 85, 88, 89, 90, 99
Cumberland, 59
Cumbernauld, 19, 47, 49
Cumbria, 56
Cunningham, 68, 72
cwm, 99

dá, 20
Da Biggins, 122, 123
dabhach, 42, 50
Dail na bhFàd, 27
dales, 99
Dalgarnock, 38
Dalmeny, 31
Dalnagairn, 27
Dalnavert, 27
dalr, 85, 99
Damnagclaur, 29
Daugleddau, 106
deans, 99
Deer, 24, 33, 35, 36, 46, 49, 50, 52
Degsastán, 67
Deheubarth, 178
Demetae, 115
Denbigh, 103
Denmark, 120, 121, 136, 137, 138
denu, 85, 88, 89, 99
Derreen, 33
Derwydd, 173, 178
Deuchrie Dod, 65
Devon, 76, 85, 89
din, 103, 105, 106
dinas, 103, 105, 106
Doarlish Cashen, 122
Donegal, 18, 26, 46
Doon Hill, 65
Dorset, 77
Drimore, 122
Dromin, 33
druim, 4, 19
Drumgoudrom, 19, 27, 29
Drummuddioch, 30, 50
Drumnanaliv, 25
Dumfriesshire, 56, 72
dūn, 76, 77, 78, 79, 98

Index

dùn, 5, 19, 103
Dunbar, 67
Dunbartonshire, 38, 50
Dun-da-gu, 20
Dundalav, 20
Dundarave, 20
Dundee, 19, 103
Dundrod, 25
Dundurcus, 19, 27, 29
Dunfermline, 103
Dunlappie, 38
Dunman, 30, 50
Dunnamaggan, 25
Dunnichen Hill, 67
Dunrossness, 140
Dunshelt, 8
Dunsinnan Hill, 67
Dunveoch, 29, 50
Dyfed, 62, 102, 103, 105, 106, 108, 109, 111, 112, 115, 117, 169, 171

eabar, 4
Eaglesham, 68
Earith, 95, 100
Eas Ruaidh, 158, 159, 161
Easbag, 20
East Renfrewshire, 68
Edinburgh, 44, 67, 70, 121
Edlesborough, 78, 80, 98
ēg, 91, 92, 93
eglwys, 111, 112, 113
Eglwys Cyffig, 111
Eglwys Ddiflodau, 113
Eglwys Fair a Churig, 111
Eglwys Gymyn, 111
Eglwys Newydd, 112
Eglwys Wrw, 111
eið, 139
Eilean Mhain, 164
Encombe, 90, 99
Englishries, 108
Eskdalemuir, 56

fàn, 21, 48
Faroe Islands, 123, 135
Fawler, 5
fell, 72
fen, 93, 94, 100

Fenlands, 92
Fens, The, 94
Fereneze Hills, 70
Fetteresso, 40, 42
Fetternear, 49
Fife, 3, 8
Finegand, 27
Fogo, 52
Foilnaman, 25
Forth, 62, 65, 68
Fort William, 65
Framgord, 130, 141
France, 35, 159
fraoch, 151, 162, 165
fraoch eilean, 162
Freswick, 121
Fulvens, 94, 100

Galloway, 29, 30, 50
gaoth, 4, 17, 47
Gaoth Beara, 47
Gaoth Dobhair, 47
Gaoth Sáile, 47
Gardie, 128, 141
Gardie I (Brookpoint), 126, 129
Gardie II (Soterberg), 126, 129
garðr, 141
Garioch, 38
Garn, 176
Garsington, 78, 79, 98
gasg, 4
gerði, 141
Germany, 159
Gilchriston, 49
gill, 72
Gillequdberit, 68
Gilmerton, 49
giolla-, 27
glac, 4
Glamis, 62
Gleann, 19
Gleann Cuaich, 165
Gleann Síodh, 154, 165, 166
Glen Noe, 19
Glenageary, 25
Glendaruel, 20
Glenshee, 154, 156, 165, 166
Glyn Farm, 173

Glyn Henllan, 173
Glyn-yr-Henllan, 173, 175, 176
gó, 47
Gogy, 38
gorsedd, 176
Gorslas, 169
Gorteen, 33
Gorthie, 38
grain, 72
Gran(d)borough, 78, 80, 98
Grange, 21
Greenland, 135
Greenock, 51
*gronn, 4
gulba, 154
gwaun, 173, 175
Gwaun Henllan, 169–79
Gweebarra, 47
Gweedore, 47
Gweesalia, 47
gwely, 177, 178

Haddo, 40
hafod, 62
hafoty, 62
Hamar, 126, 127, 128, 129, 135, 139, 140
Hannigarth, 141
Haroldswick, 123, 125, 128, 140, 141
Haselor, 84, 99
Hatton Hill, 67
Hayston Hill, 67
Hedon, 78
Henderston Hill, 67
Hendre Gored, 175, 176
henllan, 173, 175
Herefordshire, 5
Hill of Alyth, 67
Hill of Drimmie, 67
Hill of Keillor, 67
Hill of Lour, 67
Hill of Loyal, 67
Hill of St Fink, 67
hōh, 81, 83, 98
hóll, 140
holm, 116
holmr, 100
hop, 90, 91, 92, 99
Hope Dale, 91, 99

Houlligarth, 141
Housigarth, 141
Hudspeth, 97, 100
Humbie, 70
hȳth, 94, 95, 96, 100

Iceland, 135
Ilanaguy, 30
inbhear, 19, 56
Inchture, 21
Ingliston Hill, 67
Ingoe, 81, 83, 99
innis, 116
inver-, see *inbhear*
Invernauld, 19, 47, 49
Iona, 162
Iscennen, 176
Islay, 162
Isle of Man, 47, 50, 122
Ivinghoe, 81, 83, 98

Jarlshof, 122, 123, 135, 142
Jura, 162

kálfr, 141
Keillor Hill, 67
Kelso, 52
Kent, 77
Kersie, 31
Kidwelly, 176
Kilbride, 49
Kilbroney, 22
Kilcalmkill, 26
Kilchrist, 26
Kilcomin, 33
Kilfeaghan, 23
Killeen, 33
Killiehangie, 26
Kilmundie, 65
Kilpatrick, 23
Kilpeter, 26
Kilpheder, 23, 122
Kilrymont, 3
Kiltarlity, 26
Kinaldy, 31
Kinclune Hill, 67
King's Seat, 67
Kinglassie, 31

Index

Kingoldrum, 47
Kingsey, 92, 100
Kingsmuir, 3
Kinminity, 65
Kinmonth, 65
Kinmount, 65
Kinmundy, 65
Kinpurney Hill, 67
Knapperna, 52
Knockman, 30, 50
Knocknamad, 30, 50
Knocknavar, 30
Knockvadie, 23
Kvívík, 135
Kyle, 68, 72

Lagnimawn, 30, 50
làirig, 4
lamh, 20
lammas, 116
Lammer Law, 65
Lammermuir, 67, 70
Lammermuir Edge, 65
Lan, 113, 114
Lancaster, 176
lanerc, 55
Langbaurgh, 81
Langside, 8
larach tigh Meidhe, 164
Largo, 38
Laugharne, 112
Leaston, 70
Leinster, 17, 18
Leitir Móir, 47
Lenabo, 50
Lewknor, 85
Lichfield, 171
limekiln, 5
Lindsey, 67
lios, 5
Littlehamar, 139
llan, 111, 112, 113
Llandeilo Fawr, 171
Llandybïe, 9, 169, 170, 171, 172, 173, 174, 175, 176, 177, 178
Llanfihangel Aberbythych, 169, 173, 175
Llangain, 112
Llangattock, 62

Llangynog, 111
Llan-newydd, 112
Llawhaden, 109
llys, 173
Loch, 19
Loch an Tuirc, 165
Loch Awe, 162
Loch Baah, 151
Loch Fraochaidh, 162, 164, 165
Loch Freuchie, 162, 163, 164, 165
Loch Máigh, 151, 164
Loch Neldricken, 19
Loch Nell, 19
Loch Ness, 62
Loch of Belmont, 142
Loch Venachar, 154
Lochan na gCat, 49
Loch-Cuaich, 165
Lochlann, 159
Lochnaw, 19
Lochnawean, 19
Logie, 31
Loinveg, 27
longphort, 5
Lothian, 56, 62, 70, 72
Lough Neagh, 19
Loughguile, 19
Lownie Hill, 67
Lurgyndaspok, 21

maenor, 173, 178
Maenor Meddynfych, 171, 172, 178
Magh, 19
Maidenhead, 94
mainaur med diminih, 171
Maybole, 68
meadar, 28
Meddynfych, 173
Melrose, 67
Mendick Hill, 65
Middlezoy, 92, 100
*mig, 4
Milnathort, 5
Mindrum, 65
minit, 59, 62
Minto, 52
Mitford, 97
mòine, 3, 27, 28

monadh, 3, 27, 28
Monadh Liath, 65
Monadh Mòr, 65
Monadh Ruadh, 65
Monaltrie, 31
Moness, 26
moneth, 62
Monzie, 49
Monzievaird, 49
mōr, 93, 100
mór, 47
Moray Firth, 59, 65
Morpeth, 97, 100
Mount, The, 65
Mounth, 56, 65
Mountlothian, 65
Moygashel, 19
Moynalty, 19
Muckairn, 26, 27, 28, 29
Mucomir, 26, 29
Muileann, 19
Muine, 27
Mull, 20, 165
Mullingar, 19
Mundie, 65
munet, 59, 62
Munlochy, 26, 27
Munro, 49
mynwent, 113
mynydd, 62
Mynydd Castlebythe, 62
Mynydd Du, 62
Mynydd Llangatwg, 62
Mynydd Llangorse, 62
Mynyddfawr, 62, 65

neimhidh, 47, 56
nemeton, 3, 56, 58
Ness of Wadbister, 142
Neubotle, 69
Nevay, 56, 59
Nevie, 56
Newbie, 70
Newchurch, 112
Newlands Hill, 65
Newtibber, 56, 59
Newtyle, 56, 59
Newtyle Hill, 67

Nithsdale, 59, 68
Norhamar, 139
North America, 135
Northmavine, 139
Northumberland, 59, 65, 67, 77, 81, 97
Northumbria, 68
Norway, 121, 125, 134, 136, 159

Ochiltree, 62, 63
*ofer, 81, 84, 85, 87, 88, 99
ōfer, 81
Ogilface, 62, 63
Ogilvies, 62
ōra, 81, 85, 87, 88, 99
Orkney, 54, 122, 134, 135, 140, 142
Othery, 92, 93, 100
Otmoor, 93, 100
Over, 81
Oxendean, 89, 99
Oxfordshire, 77, 78, 92
Oxnop, 99
øy, 141

pæth, 97, 100
Papa Stour, 122
parc, 117, 118
Parc y Fynwent, 113
pebyll, 62, 64
peit, 28; see also *pett*
Pembroke, 106, 112
Pembrokeshire, 102, 103, 106, 107, 108, 113, 115, 116, 117
Penfynydd, 62, 65
Penmynydd, 62, 65
Penshiel Hill, 65
Pershore, 85, 86, 99
pert, 55
Perthshire, 24, 48, 49, 59, 65, 67, 149, 154, 162, 164, 166, 167
Petenderleyn, 20, 21
pett, 55; see also *peit*
pevr, 55
Pinkerton Hill, 65
Piode, 178
Piode fawr, 177
Pit-, see *peit*, *pett*
Pitcairn, 6
Pitcorthie, 5

Index

Pitcruive, 23
Pitforthie, 5
Pitlochie, 31
Pitsligo, 38, 39
Pogbie, 70
pōl, 59
pol, 59, 60, 61
poll, 59
Pont Cowin, 115
pow, 59
Prestwick, 68
Pusk, 68

Rainow, 81, 99
Raiss, 70
Rameldry, 6
ràmh, 20
Ramornie Mains, 6
Rannoch, 38
rath, 106, 107, 108, 116
ràth, 5, 6, 106
Rathcroghan, 162
Rathelpie, 36
Ravensby, 72
Renanirree, 25
rhandir, 176
rhandiroedd, 177
rhath, 106
Rhos, 106
riasg, 4
River Avon, 85
River Axe, 94
River Cywyn, 115
River Severn, 93, 94
River Wansbeck, 97
Roberton, 21
Rook Barugh, 78, 98
Rosdanean, 20
Roseberry, 81, 98
Roseberry Topping, 78, 80, 81
Rosneath, 47
Ross of Mull, 162
Rosser, 176
Ruadhail, 20

sætr, 140
St Andrews, 1, 3, 10
St Cynog, 111

St Dogmaels, 113
Sand, 138, 139, 142
sandr, 140
Sandwick, 130, 131, 134, 140, 141
Sandwick-North, 126, 130–4
Sandwick-South, 122, 123, 126
sarn, 6, 115, 116
Sarn Helen, 115
Scaleber, 78, 98
scir, 68
Scleofgarmonth, 65
Scoonie Hill, 3
setberg, 140
Setters, 126, 129, 140, 141
Shankill, 23, 49
Shotover, 84, 99
Sidlaw Hills, 59, 67
Skelpie, 37
Skyber Farm, 62
Skye, 166
sliabh, 19, 56, 74
Slieve Gullion, 19
Slievenamon, 25, 50
slochd, 4
Smeaton, 70
Smirgarth, 141
Solway, 56
Somerset, 92, 94
Sorbie, 72
Soterberg, 129, 140
Soutra Hill, 65
Spain, 159
Spott Dod, 65
Stainmore, 56
Standen, 89, 99
Stanhope, 91, 92, 99
Stobo, 52
Stonehaven, 65
Stoora Toft, 129, 140
Stoora Tofta, Westings, 126
stórr, 140
Stottesdon, 78, 79, 98
Stracathro, 39
Strathclyde, 56
Strathmore, 67
Suidhe, 164

Tamnaverie, 31

Index

Tanderagee, 17, 18
Tandragee, 17, 18
tangi, 140
Tara, 158
Tarvit, 21
Tay, 59, 62, 65
Temple MLO, 73
Tenby, 103
Thames, 92, 94
threep, 68, 71
Thundergay, 17
Tir Dafydd, 169, 175, 177
Tir Rosser, 176
Toftanes, 135
tóin, 17
Tóin re Gó, 47
Tonlegee, 17, 18, 47
Tonragee, 17
topt, toft, 140
Torr na Gaoith, 47
Torry, 31
Trossachs, 154, 165
Tullaherin, 23
Tulloch, 38
Tullyvin, 23
Tulrahan, 23
Turnberry, 68
Tysoe, 81, 83, 99
Tywi, 117
Tywi valley, 171, 178

uchel faes, 62
ucheldref, 62
uchelfa, 62
Udal, 122
Ulster, 17, 18, 49, 148, 149

Underhoull, 122, 123, 126, 139, 140
undir, 140
Unst, 6, 120–46
Uyeasound, 123, 140, 141

vágr, 139
Vale of Aylesbury, 77
vatn, 141, 142
Vatnsgarth, 141
View Edge, 83
vík, 140, 141, 142
Voe, 139
Votadini, 62

Wadbister, 129, 141, 142
Wæsse, 93, 100
waulkmiln, 5
waun, 173, 175
Weddersbie, 72
Wellingore, 84, 99
Welshries, 106, 173
Wentnor, 84, 99
Weo, 81, 83, 99
Westing, 129, 140
Westmorland, 59, 77
Westness on Rousay, 122
Whalsay, 140
Whitcombe, 90, 99
Widdecombe, 89
Wiston, 56, 106
wrae, 72
Wreaton, 72

Yell, 142
ynys, 116, 117, 118
Ysgubor Mountain, 62